Moussa Hamadou Ousseini

Impact of the security crisis

Moussa Hamadou Ousseini

Impact of the security crisis

Impacts of the security crisis and flooding on
bell pepper production in eastern Niger

ScienciaScripts

Imprint

Any brand names and product names mentioned in this book are subject to trademark, brand or patent protection and are trademarks or registered trademarks of their respective holders. The use of brand names, product names, common names, trade names, product descriptions etc. even without a particular marking in this work is in no way to be construed to mean that such names may be regarded as unrestricted in respect of trademark and brand protection legislation and could thus be used by anyone.

Cover image: www.ingimage.com

This book is a translation from the original published under ISBN 978-620-6-71840-6.

Publisher:
Sciencia Scripts
is a trademark of
Dodo Books Indian Ocean Ltd. and OmniScriptum S.R.L publishing group

120 High Road, East Finchley, London, N2 9ED, United Kingdom
Str. Armeneasca 28/1, office 1, Chisinau MD-2012, Republic of Moldova, Europe
Printed at: see last page
ISBN: 978-620-7-93661-8

CONTENTS

DEDICACES

I dedicate this memoir

To my parents MOUSSA Hamadou, FATI Boukar

To my twin brother who died when we were children may his soul rest in peace Ameen*!!!!*

To my brothers and sisters.

ACKNOWLEDGEMENTS

At the end of this work, we would like to express our deep gratitude to all the individuals and legal entities who have contributed to its completion.

I would like to extend my warmest thanks to Dr SOULEY Kabirou, Senior Lecturer in the Geography Department of the Faculty of Arts and Humanities at the André Salifou University in Zinder, who, despite his many concerns, agreed to supervise this work.
Our thanks also go to Dr ABDOU BAGNA Amadou, Assistant Professor in the Geography Department of the Ecole Normale Supérieure at the Abdou Moumouni University in Niamey, who co-directed this work. I would like to express my deep gratitude to all of them.

We would like to express our sincere gratitude to all the teachers in the Geography Department of the André Salifou University in Zinder who have supported us throughout our university studies.
I would like to thank Dr ILLOU Mahamadou, Senior Lecturer and Research Teacher at the André Salifou University in Zinder in the Geography Department, who chaired the defence of this dissertation.
I would also like to thank Dr ABDOU Rabiou, Maitre-Assistant and Enseignant Chercheur at the André Salifou University of Zinder in the Faculty of Science and Technology, who agreed to act as assessor in order to improve the quality of this document.

ABDOU Mahaman for his support throughout my stay in Zinder. I would like to express my sincere thanks to him for his advice and financial assistance.

We would also like to thank all our classmates, in particular Mr ABDOU SANI Mountaka, Mr MOUTA ADAM Lawali, Mr ABDOU ALI Atahir, MAHAMADOU MOUDI Rachid and YACOUBA ISSOUFOU Achirou.

My friends and acquaintances for their sympathy and profound kindness who have participated in one way or another in the realisation of this work, my sincere thanks and know in this work, the fruit of your efforts.

Thank you all !!!!!

SUMMARY

This study looks at the impact of the security crisis and flooding on pepper production in the Commune Urbaine of Diffa. The effects of the security crisis are a real tragedy for the population in general and farmers in particular. The damage caused to the region's economy is considerable. The overall aim of the study is to show the impact of the security crisis and flooding on pepper production in the Diffa Urban Commune. The methodological tools used were documentary research, direct field observation, data collection through quantitative and qualitative surveys, and finally the processing and analysis of the data collected. The results show that the security crisis and recurrent flooding have had a negative impact on the pepper industry in the urban commune of Diffa. First of all, 45% of the producers surveyed have stopped pepper production because of the Boko Haram insecurity and 32% have abandoned their production land because of the recurrent flooding. The reduction in the area under cultivation and the fall in production were the most significant impacts, resulting respectively in a reduction of almost 78.4 hectares, or 73.17% of the area under cultivation, and a fall in production of around 85.11%. As a result, the farmers affected are living in a precarious situation. Farmers have developed local strategies to cope with these new situations.

Key words: Diffa Urban Community, pepper production, security crisis, flooding, impacts.

GENERAL INTRODUCTION

Niger is a landlocked country covering 1,267,000 km², with its southernmost border more han 600 km from the sea (Gulf of Guinea). Three quarters of its surface area is located in the Saharan zone. The climate is tropical, semi-arid to arid. The rainfall pattern is characterised by low rainfall that varies in space and time, and high temperatures that tend to accentuate the aridity of the climate. MINISTRY OF AGRICULTURAL DEVELOPMENT, (2006).

Demographically, with a natural growth rate of 3.9% (the highest in the world), its populat on is estimated to be 25,112,469 in 2022[1] , over 80% of whom live in rural areas INS, (2012). This strong demographic growth is increasing pressure on natural resources and exacerbating the vulnerability of populations, particularly in rural areas (PNEDD, 2000).

Niger's economy is largely based on the primary sector. In 2017, the primary sector accounted for 43.4% of GDP (INS, 2018). The contribution of irrigated agriculture export earnings from irrigated crops (especially onions and peppers) exceed CFAF 10 billion[2] MINISTRY OF AGRICULTURE, (2015). Niger has an estimated irrigable potential of between 270,000 ard 330,000 hectares[3] , i.e. between 1.8% and 2.2% of the total arable land in the onion, souchet, sesame and pepper sectors PDES (2015).

The practice of irrigated farming only became widespread in Niger after the major droughts of 1974 and 1984. Since then, the government of Niger has continued to support the policy of irrigated farming throughout the country. These crops are grown around the various wate: points - boreholes, wells, cesspools, ponds, lakes and rivers (OUMAROU H, 2005). Indeed, the main crop cultivated via irrigation in the Diffa Region, pepper growing, remains the main activity of the local population. In normal times, production is estimated at around 11,00C tonnes of peppers per year and brings in around 6 to 10 billion CFA francs. In 2014, production was estimated at 11,142,820 tonnes, or around 655,460 bags. The revenue estimated by the CRA is 19,663,800,000 CFA francs. Given its importance to the household economy, the pepper is nicknamed "the red gold of *Manga*" PAM, (2016).

Peppers, with their considerable contribution, represent a promising niche market for the Diffa region and for the country as a whole. The Komadougou is the pepper production zone. This area is under-exploiting its considerable potential for development, estimated at 75,000

---

[1] The law of geometric increases has been used to estimate the population in 2022 $P = P_0 (1+T)^n$

[2] SNDI/CER, 2005 and the study on the contribution of irrigated agriculture to agricultural GDP.

[3] SOGEAH/BRGM inventory in November 1981. Data on the national potential for developable land have not been updated since 1982.

hectares. In fact, only 8,000 ha are currently developed for market garden crops and cash crops, mainly peppers, okra, onions and wheat (ABDOU A et *al,* 2005).

Since 2013, the conflict linked to the *Boko Haram* sect in the Lake Chad Basin has caused the displacement of thousands of refugees, IDPs and returnees forced to give up the right to access land and agricultural assets UNHCR, (2017).

Flooding is now at the heart of the disaster risks facing Niger. The country has been facing recurrent flooding for several years (HAMANI O and ABDOURAHAMANE O, 2017). Natural and anthropogenic factors are at the root of the recurrent floods in the Urban Commune of Diffa, the repercussions of which have been significant in recent years.

The security situation and recurrent flooding in Diffa prompted the choice of the theme: *"Impacts of the security crisis and flooding on pepper production in the Urban Commune of Diffa".*

In addition to the general introduction, the discussion and the general conclusion, this work is structured into four (4) chapters. The first chapter deals with the theoretical and methodological framework, the second chapter presents the general characteristics of the study area, the third chapter describes the pepper production system in the study area and the fourth chapter analyses the impact of the security crisis and flooding on pepper production and the local strategies for dealing with them.

CHAPTER I: THEORETICAL AND METHODOLOGICAL FRAMEWORK

Introduction

This chapter describes the rationale behind the choice of topic, the issues raised by the subject and the literature review. It then presents the hypotheses, objectives and definitions of a number of key concepts. It also discusses the appropriate methodology. Finally, it details the difficulties encountered in carrying out this work.

1.1 Theoretical framework

1.1.1 Justification for the choice of theme.

This topic is of great importance to our study area for two reasons: the security crisis and flooding in the Manga capital, and the importance of pepper growing.

Flooding has been the talk of the Commune Urbaine de Diffa for the past decade. This natural disaster continues to occur almost every year, causing extensive damage.

In Niger, any mention of peppers automatically refers to the Diffa Region. More than 80% of national pepper production comes from the Diffa Region (MALIKI et *al*, 2006). Given this state of affairs, Diffa is a favourable area for carrying out this study. The study area has enormous potential for irrigated cultivation, including :

➢ The existence of vast tracts of irrigable land, only 4% of which is developed;

➢ Availability of surface and ground water ;

➢ A dynamic pepper sector;

➢ The soil is relatively fertile.

The issue of the security crisis and its socio-economic impact is a recent phenomenon that deserves to be studied in this Region. The aim of this study is to highlight the impact of the security crisis and flooding on pepper growers. This research is a modest contribution to the many studies already carried out on the problems of insecurity and its impact on agriculture in this area, including :

➢ OXFAM

➢ CRA

In addition, this study will enable us to put into practice what we have learnt during our geography training, by designing and carrying out research work.

1.1.2 Issues

Since 2009, the Lake Chad region, which comprises the four riparian countries of Niger, Nigeria, Cameroon and Chad, has been experiencing an unprecedented situation, characterised by an unprecedented security crisis. The Diffa region, which shares the vulnerability of the Sahel, is subject to strong demographic growth, growing insecurity, political instability and the adverse effects of climate change. The Lake Chad region has a predominantly rural population, and its economy is essentially based on agriculture, livestock farming and fishing, on which rural people depend. (GERAUD M, and MARC-ANTOINE P, 2018)[4] . The armed insurrection of the *Boko Haram* group and its repression from 2009 onwards are now one of the major crisis areas on the African continent, with its share of victims, hundreds of thousands of displaced persons and the spectre of famine. In Nigeria alone, the country hardest hit by the conflict, the World Bank estimates the damage at billions of dollars in 2015 WORLD BANK, (2015)[5] . Insecurity and the measures put in place by counter-insurgency actors have deserted certain areas. Yet these are often areas that represented the most productive rural hubs, namely Lake Chad and the Komadougou Yobé (GERAUD M, and MARC-ANTOINE P, 2018).

Like the other countries in the Lake Chad basin, Niger has been facing serious security threats along its borders for the past decade and a half. In fact, the north of Niger is a route for jihadist groups operating in northern Mali and southern Algeria. In the south-east, Niger is under direct threat from armed groups from *Boko Haram*, who have made several incursions into the Diffa region. In the north-west, the security situation in northern Mali, particularly in the Kidal region and in the Niger loop, is far from stabilised. Indeed, in general, security conditions have deteriorated further in the border areas with Mali, Burkina Faso and north-eastern Nigeria MINISTRE DU PLAN, (2017).

According to OCHA (2020), there are four (4) fronts to the security crisis in Niger.

> ➢ Since 2012, the western front along the north-western border, insecurity in Mali and Burkina Faso and repeated incursions by non-state armed groups (GANE) in the Tillabéry and Tahoua regions have had an impact on the living conditions of the people of Niger.

> ➢ The eastern front in the south-east, in the Diffa region, where the security situation is still marked by repeated attacks by the GANE (some 674 people were killed, wounded or abducted by the jihadists between January and August 2019, leading to secondary movements of people seeking protection in the Communes of Diffa and Gueskérou. As a reminder, the state of emergency in the region has been in force since 11 February 2015, following simultaneous attacks by *Boko Haram* on the towns of Diffa and Bosso on 06 February 2015. Some 260,000 people (displaced persons, refugees from Nigeria and returnees) have been registered in the Diffa region, forced to flee on numerous occasions.

---

[4] Find out more about our publications: http://editions.afd.fr/
[5] According to the World Bank's Recovery and Peace Building Assessment (RPBA) (2015), the fighting caused cumulative production losses of 8.3 billion US dollars (USD), destroyed 9.2 billion USD worth of infrastructure, killed around 20,000 people and displaced 1.8 million.

➤ In southern Niger, the deteriorating security situation on the border with Nigeria has also led to new movements of populations who have settled in the border areas of the Maradi Region. Indeed, since May 2019, more than 35,000 Nigerians from the states of Sokoto, Zamfara and Katsina have arrived in the Maradi Region (OCHA, 2020).

➤ In the north, the Libyan conflict has spread to the desert area bordering Libya. This area has become the main centre for arms and drug trafficking by criminal groups from Libya.

An agropastoral country par excellence, three-quarters of Niger's surface area is located in the Saharan zone. The climate is tropical, semi-arid to arid. The rainfall-temperature regime is characterised by low rainfall that varies in space and time, and high temperatures that tend to accentuate the aridity of the climate. Despite these natural constraints, agriculture is practised by more than 80% of the population. Agriculture (including livestock, fisheries and forestry) accounted for 42.1% of GDP in 2014 MINISTERE DU PLAN, (2017). In addition to the drought, demographic pressure is illustrated by a population estimated to be 25,112,469 in 2022, with one of the highest annual population growth rates in the world, i.e. 3.9%, INS, (2012). This demographic dynamism increases the pressure on natural resources and worsens the vulnerability of populations, particularly in rural areas, PNEDD, (2000). After the major droughts of 1968 and 1973, the government of Niger introduced irrigated farming throughout the country to deal with the food insecurity that had become almost permanent (COHAND L, 2007).

As in other parts of the country, desertification and its effects are a major problem in the Diffa region, undermining its viability. Rainfall during the main agricultural production season decreases progressively from the south to the north of the region, giving it a pastoral character with extensive grazing (ADAMOU M, 2019). This region is facing a security situation marked mainly by repeated attacks by *Boko haram*. This security crisis has resulted in massive population displacements and inaccessibility to agro-sylvo-pastoral and fisheries production areas (PAM, 2016).

This situation of insecurity has led to the destabilisation of the production and marketing of peppers and fish, which are the main sources of income for the people of the region, particularly along the Komadougou River and the Lake Chad basin. Agriculture and livestock farming are the two (02) mainstays of the Diffa Region's economy. As a result, they are subject to the hazards that inevitably affect yields. The main food crops are millet, sorghum, cowpea and rice, whose production does not cover the needs of the population. The main cash crop is peppers, whose production is subject to the vagaries of the weather PDC, (2014). The region receives very little rainfall due to its location in the Saharan and Sahelian zones. However, like the rest of the country, agriculture is facing a sharp deterioration in its production potential, under the combined effect of increasing land pressure, irregular and unfavourable climatic conditions, and production methods that remain traditional. All these factors, combined with the nature of the production systems, have resulted in a chronic cereals deficit. PIERRE-FRANÇOIS P and SALIFOU K, (2005).

Located in the Diffa Region, the Commune Urbaine of Diffa extends over a radius of 20 km on either side of the urban centre, with an estimated surface area of 229 km2. With a population of 78,544 in 2022, it is predominantly rural.

The economic fabric of our study area is dominated by the primary sector, i.e. agriculture, livestock farming and fishing. However, these sectors are threatened by the very restrictive conditions of insufficient and irregular rainfall and insecurity due to threats from the *Boko-Haram* sect, which have caused a massive displacement of people from the Komadougou Yobé valley to the north. This displacement has led to the abandonment of irrigated crops. Peppers, the region's "red gold", are the main cash crop. The cultivation of sweet peppers appears to be a strategic choice to allocate available resources to a secure irrigated cash crop, in a context where the success of rain-fed crops is uncertain. However, the major constraint of this sector is the difficulty of access to finance faced by all players, and more particularly pepper growers (PIERRE-FRANÇOIS P and SALIFOU K, 2005).

In fact, the weakness of the organisation of the players is holding back the development of the sector. There are major threats to the sustainability and profitability of production systems due to cultivation practices that lead to soil impoverishment, lower profitability, harmful effects on the environment and insecurity in the area. (PIERRE-FRANÇOIS P and SALIFOU K, 2005). In addition to all these constraints, there is the recurrent flooding caused by the rising waters of the Komadougou Yobé. The years in question were 1998, 1999, 2002, 2012, 2013 and 2019, of which 2019 was an exceptional and catastrophic event due to the scale of the losses and damage caused to the population.

Generally speaking, the security crisis, flooding and pepper production are therefore important issues for the population of the Diffa Region. This justifies this research work on the theme "Impacts of the security crisis and flooding on pepper production". To this end, the following main question is posed: What are the impacts of the security crisis and flooding on pepper production in the Urban Commune of Diffa? This main research question gave rise to other secondary questions that served as a basis for reflection:

➤ What does the pepper growing system look like?

> What impact have the security crisis and flooding had on pepper production in the Commune Urbaine de Diffa?

> What endogenous strategies have farmers adopted to cope with the various constraints, particularly insecurity and flooding?

1.1.3 Literature review

Much work has been devoted to studying the security crisis, its effects and the adaptation strategies adopted to deal with it. This literature review is divided into three sections: the context of the security crisis, recurrent flooding and pepper growing.

> **The security crisis, a brake on economic activity in the region**

In the Sahel since 2010, violence linked to Islamist militias has increased considerably on the continent, accounting for 13.54% of all political violence in 2012 (compared with 4.96% in 1997) and in West Africa, extremist groups now move easily from one state to another according to (STRAUS, 2012 quoted by ALEXANDRE M, *al*, 2015).

In this context, OCHA (2017) states that in the Lake Chad Basin, as of March 2017, the security crisis had caused the displacement of almost 2.5 million internally displaced people, refugees and returnees. This makes it the second largest displacement crisis in the world after Syria, and the fastest growing. In the last two years, the number of displaced people, 58% of whom are women and 62% children, has tripled.

In its assessment, the FAO (2017) highlights the significant loss of means of production, the inaccessibility of production sites and the destruction of marketing channels for the supply of inputs.

In particular, the spread of the *Boko Haram* phenomenon to the lake and the Chadian shores still entails security risks and has the potential to destabilise a cosmopolitan region in which the mechanisms for cohabitation remain fragile. In addition to seasonal movements, the successive arrivals of displaced civilians, refugees or returnees on the shores of the lake since January 2015 have sometimes affected social cohesion. In January 2017, the humanitarian agencies OCHA and OXFAM reported the presence of 100,765 displaced people in the region, to which should be added almost 21,000 displaced people not yet registered and almost 7,000 refugees, mainly Nigerians. For the most part, they have fled the atrocities of *Boko Haram* or left the islands under pressure from the region's armies INTERNATIONAL CRISIS GROUP RAPPORT, (2017).

In addition, according to a paper by (SAGAGI M and THORBUM A, 2019), the Lake Chad region is frequently cited as a region suffering from the consequences of climate change and desertification, resulting in limited economic activities around agriculture, fisheries and trade, which in turn increase susceptibility to violence and crime. The four countries of the Lake Chad region are characterised by populations that are mostly poor and depend primarily on subsistence agriculture, with limited opportunities for value addition and innovation.

As the work of UN, (2018) has revealed, the state of emergency put in place in 2015 in the Diffa region, and renewed several times, has had the effect of strengthening the Nigerien military presence and the multinational special force, made up of military units from Benin, Cameroon, Niger, Nigeria and Chad, with a mandate to put an end to the conflict started by *Boko Haram*. In addition, the prolonged conflict and the measures adopted under the state of emergency have limited livelihood activities, including fishing, fish sales, pepper production, and the purchase of fuel and fertilisers essential for agricultural production. The state of emergency has also led to restrictions on people's freedom of movement, with areas militarised and declared inaccessible to civilians. These measures, aimed at combating non-state armed groups, have had a devastating effect on the civilian population and their livelihoods in the Diffa Region.

Indeed, the study conducted by the FAO, (2016) in Borno, Yobe and Adamawa states in Nigeria showed that most farmers in the past had a fairly resilient informal seed system, with strong social networks and multiple varieties grown by each household. Since the *Boko Haram* insurgency, both formal and informal seed systems have been disrupted. Certified seed from formal sources is reportedly too expensive for many people, particularly poor farming households and internally displaced people, refugees and returnees. Poor roads and insecurity increase transport costs, leading to higher market prices.

Also OXFAM-Diffa, (2016) points out that since February 2015, the value and volume of the dried red pepper market in the Diffa region has fallen sharply. The Komadougou area has become heavily militarised while recording a huge increase in refugees and displaced people in 2015 and 2016.

The author explains that many pepper growers have found themselves displaced and separated from their productive land. As for those who have remained close to their land, most have only limited access to it, at great risk, as the farmers interviewed by Oxfam in 2016 pointed out. Oxfam's data shows a 96% drop in the number of people claiming to earn an income from pepper cultivation since the start of the conflict. Those who continue to work as small-scale pepper farmers earn only 64% of what they did before the *Boko Haram* conflict began. Their production has fallen from 100 to 50 bags per pepper harvest to an average of 50 to 10. In addition, people are very fearful of the attacks, looting and extortion perpetrated by *Boko Haram* in the fields, on the roads and in their villages, especially at night. In addition, because of the strong military presence, some farmers, mainly women, do not feel safe accessing their fields. Under the current state of emergency, pepper growers are required to obtain authorisation to purchase agricultural inputs, particularly fertilisers and fuel, which the government says can be used to make explosives. As a result of these rules, growers have limited quantities of fertiliser and are therefore unable to fertilise the same areas of land as before. These restrictive measures also tend to force people to engage in illicit trade to gain access to the necessary inputs, which increases the risks of continuing their activities. The incomes of small pepper growers have collapsed as a result of low prices and reduced production capacity, mainly due to restrictions on access to productive land OXFAM, (2016). In this context, the UN (2016) has estimated that nearly 8 million people in the Lake Chad region are in need of assistance. In Niger, for example, the Diffa camp, run by the United Nations, hosted around 250,000 displaced people in 2016. That year, *Boko Haram* carried out more than 30 deadly attacks on the camp, forcing the UN to relocate these refugees to a neighbouring area. Since then, although reports indicate that the security situation has improved in Diffa, the Lake Chad Basin is suffering the consequences of *Boko Haram*'s actions in the EASO region, (2018).

> **Background to recurrent flooding**
> - **Flooding as a source of natural disasters**

Floods are the world's biggest natural disaster, claiming around 20,000 victims a year (MEDD, 2004). They are among the most common natural disasters in the world, and the most costly in socio-economic terms.

Flooding is one of the most frequent and devastating natural disasters, affecting several regions and sometimes reaching the scale of a national disaster (BOUDOUKHA A, and BOUMESSNEGH A, 2012). Flooding has become synonymous with loss and devastation. Almost every year, we learn that somewhere floods have caused damage to infrastructure, livelihoods and property.

According to WMO, (2006) quoted by BENOIT S, (undated) more than 80 to 90% of natural disasters are linked to hydro-climatic events such as droughts, heavy rain and floods.

- **The consequences of flooding**

For some decades now, flooding has been one of the most feared consequences of extreme climatic variability. It is one of the most devastating scourges to hit the planet today. It is characterised by the submergence of land resulting from the overflow of water or heavy rainfall (DOSSOU, 2015).

The consequences of flooding are both environmental and human. They result in the loss of human life and property, the engulfment of arable land and damage to economic infrastructure (AGRHYMET, 2013 Cited by HAMANI O and ABDOURAHAMANE O, 2017).

After the droughts of the 70s and 80s, the countries of West Africa, and the Sahel in particular, are now suffering the effects of heavy rains and devastating floods. The damage and losses associated with these extreme hydro-climatic events have been estimated at several hundred billion Swiss francs. In addition, these events have undermined human systems, causing human and material losses, sunken agricultural systems and economic infrastructures BENOIT S, (undated).

According to the WFP (2007), "the 2007 floods in Africa, which stretched from Mauritania in the west to Kenya in the east", are considered to be the worst in recent decades. More than one and a half million people were affected, including more than 600,000 in West Africa.

In 2007, the worst floods in West Africa for over 30 years killed 33 people in Burkina Faso and 23 in northern Togo, and displaced 46,000 people, including 26,000 in Burkina Faso and 14,000 in Togo. In Burkina Faso, 17,689 ha of crops were flooded, with production losses of around 13,500 tonnes, and 55 dams had their dykes breached (ADAPTE S, 2009).

The economic cost of flooding is often considerable. Between 2000 and 2008, flood-related damage in the CILSS region was estimated at between thirty-nine (39) and eighty (80) billion US dollars (AGRHYMET, 2013). As far as property is concerned, market-garden crops have been washed away, and communications and electricity networks have been disrupted, adding to the economic losses.

Like many West African countries, Niger is subject to the risk of flooding, which causes considerable damage. According to a study carried out in 2013 by the Institut de la Recherche en Développement (IRD, 2013), flooding in Niger has affected almost all regions to varying degrees over the past decade. These floods occur between August and November.

According to HAMADOU and DOMINIQUE (2013), one of the most serious floods to hit the capital Niamey was in 1998, when major losses were recorded. This involved the collapse of dwellings and land communication structures (roads, bridges, etc.). In addition to the physical damage, there is the psychological damage caused by trauma and social instability. Floods can also have a negative impact on education as a result of the dysfunction they can cause.

Thus according to, (HAMANI O and ABDOURAHAMANE O, 2017), mentions the balance sheet of the 2012 floods. It reports that the 2012 floods damaged socio-economic infrastructure and irrigated perimeters along the River Niger and the Komadougou and caused significant damage. In the same issue of Sahel quotidien n°8997, According to the ministry in charge of agriculture, as of 7 September 2013, the 2012 floods caused damage estimated at 32 billion CFA francs 48.8 million euros.

Like the other regions of Niger, Diffa is not spared from these phenomena. In fact, like most of the country's other regions, the Diffa Region suffers from catastrophic flooding, which is a major constraint on activities and an obstacle to economic development.

The flood inventory for the period 2000-2021, according to DRM/D data, shows that the Diffa Urban Commune has experienced repeated flooding. These unpredictable events in time and space have caused enormous damage. According to the 2019 flood report drawn up by the MAH/GC/ Diffa focal point, 1,187 households were affected, i.e. 6,717 people, 1,909 houses and 235 huts collapsed.

According to the work of TARCHIANI V, et *al* (2021), flooding in the Diffa region is concentrated in the south of the region and is mainly due to flooding of the Komadougou, a tributary of Lake Chad, or to heavy rains.

According to the study carried out by the CNEDD (2019), in the event of flooding, almost 120,600 ha of the region's farmland, consisting mainly of irrigated and lowland crops, will be at very high risk of vulnerability. These are mainly pepper and rice crops. Localities along the Komadougou River and on the shores of Lake Chad are exposed to a high risk of vulnerability to flooding (as in 1988, 2012, 2015 and 2017).

➤ **Pepper production**

• **Characteristics of peppers**

Market garden crops are of great importance to mankind, helping to combat malnutrition throughout the world, particularly in tropical areas affected by drought, and also enabling farmers to increase their sources of income (Candy J, 2006). In Côte d'Ivoire, in the localities of Dabou and Bouaké, pepper cultivation follows this principle (ANTOINE K, 2015). The sweet pepper (*Capsicum annuum*), better known by its generic name of capsicum, is a vegetable plant native to tropical America, highly prized for its fruit, which is mainly consumed as a vegetable (TRISTAN, 2004). Peppers belong to the Solanaceae family. The first pepper varieties are thought to have originated in an area located in the South Brazilian mountains to the east of Bolivia, Paraguay and the south of Argentina. This area is known as the "central area", where all the main domesticated species of the genus are thought to have been found (KALLOO, 1988 cited by ANTOINE K, 2015).

The pepper plant is a herbaceous plant, 0.5 to 1.5 m high, with a fairly strong taproot system and roots that tend to grow laterally within a radius of 0.30 to 0.50 cm. The stem gradually becomes lignified, giving rise to a tendency towards perennial growth. The leaves are simple, broad, soft, petiolate and alternate, often hairless. The *Capsicum annuum* species is the most widely cultivated in the world and the most economically viable (CHAUX and FOUR Y, 1994 cited by ANTOINE K, 2015). *Capsicum annuum* includes *sweet peppers* and prickly varieties. The world's main producers are China, followed by Mexico, Turkey and Spain. Peppers are one of the most temperature-demanding of all vegetable plants, but are less sun-demanding than tomatoes. It thrives best at temperatures of between 16° and 26°C, with 50-60% tropical sunlight, especially for young plants. In the tropics, it thrives at altitudes of between 400 and 800 metres, as well as during the dry season in Sahelian or South Chinese climates at around 25°N latitude. It is increasingly grown in the savannah during the dry season for export to Europe in the off-season. It requires humus-rich soil with low humidity (Candy J, 2006).

- **Socio-economic contribution of peppers**

Peppers are still one of the most widely grown crops worldwide, across all continents. In addition, world pepper production has evolved progressively over time, from around 20 million tonnes in 2000 to 22 million tonnes in 2013, representing an annual increase of around 4% (BELKHIRI E, and SEFIH F, 2016). However, according to the FAO (2007), world pepper production is estimated at 23.2 million tonnes, with China the world's leading producer with 14 million tonnes, i.e. almost 50%.

According to FAO statistics, pepper production in the Mediterranean basin exceeded 100 million tonnes in 2015. Turkey, Egypt, Italy and Spain account for 71% of this production. Algeria is by far the main producer, with an average annual production of 6788809 tonnes, i.e. 0.1% of total world production, and 09% of production in the Mediterranean basin (ZITOUNI D, and DOUAR K., 2017).

In this context (ISSA Y, 2005) points out that in Niger, peppers are produced mainly in the Diffa department (82% of national production)[6] and particularly in the arrondissement of the

---

[6] Agro-pastoral export promotion project, study on trade facilitation covering the PPEAP agricultural sectors (undated).

same name. Other production takes place almost everywhere in the so-called off-season sites of Zinder, Maradi and Tillabéry in particular. Producers in the south of Maradi export fresh peppers to Kano. Thus, (DIALLO B 1995 quoted by PPEAP, undated) states that national demand is estimated at 0.3 kg per inhabitant per year on the basis of 0.2 kg in rural areas and 0.8 kg in urban areas. In 1999, overall national demand was estimated at around 3,000 tonnes, or a third of production that year. Diffa peppers are exported to Nigeria, on the Maiduguri - Kano route. This is due to a difference in crop calendars between Niger (earlier) and the other side of the Nigerian border. It is estimated that around 2,000 tonnes of sweet peppers were exported to Nigeria in 1995, and the needs of this major market seem far from being met. In the case of the sweet pepper sector, net foreign currency inflows in 1998 amounted to 4.3 billion. Niger peppers are much sought after in Nigeria and production could be increased, according to market observers PPEAP (undated).

(BIGA I, 2008) states that the average yield over the last ten years is around 1,600 kg of dry pepper per hectare, compared with 2,200 kg in other regions[7] PAPAK, (2006).

With regard to pepper production in the Diffa region, (PIERRE-FRANÇOIS P and SALIFOU K, 2005) stresses that in the 1980s, the spread of small motor pumps adapted to individual farming led to a considerable increase in production and, in the 2000s, from this strip 150 km long and 5 km wide, almost 10,000 tonnes of dried pepper are exported annually, representing a windfall of 7 to 8 billion CFA francs 7 to 12 M€ benefiting a farming population estimated at more than 30,000 people.

According to the Diffa Regional Chamber of Agriculture, before 2014, the area under pepper was estimated at 8,000 ha, with dry pepper production of between 8,000 and 10,000 tonnes. Before the *Boko Haram* incursion, annual production in the sector was usually worth 15 billion CFA francs. Peppers are the "engine" crop of the Komadougou valley, providing work for thousands of households of both sexes and all age groups.

According to the Réseau National des Chambres d'Agriculture du Niger RECA (2010), an analysis of operating accounts shows that the activity is profitable for the vast majority of producers, with a commercial profitability rate of over 33% for more than half. With a high added value ratio of over 35%, the activity contributes to the creation of national wealth. However, CRA (2010) points out that operating costs are high, at around 955,000 FCFA/ha, and relate to fertilisers and pesticides (22%), labour (19%) and fuel and lubricants (15%). These

---

[7] Projet d'Appui aux Producteurs Agricoles de la Komadougou PAPAK, (2006).

high costs highlight the financing difficulties faced by farmers, forcing them to resort to informal credit.

In addition to its economic contribution to the Diffa Region, the importance of the pepper sector in food security is very high in the Diffa department, which provides most, if not all, of the production. For example, the income earned by producers in 1998 should enable them to purchase more than 8,000 tonnes of cereals, equivalent to the annual needs of more than 30,000 PPEAP people, (undated). As for (VICTORIA O, 2015)[8] capsicum, an excellent source of vitamin C, 125 ml of red capsicum provides almost double the recommended amount per day. Vitamin C is an essential antioxidant, protecting the human body from infection and promoting healthy bones, teeth, gums and cartilage.

Peppers are also used to prepare sauces for household meals or to season certain ready-to-eat foods such as fritters, grilled meat, etc. Pepper consumers therefore belong to all social strata in the country (OUSMAN Y, 2013).

From all the above, it can be seen that pepper production has been approached from a number of angles. It is adopted worldwide and makes a major contribution to improving people's living conditions through its economic contribution and social use. But in the study area, the development of this agricultural activity is constrained by climatic hazards and strong demographic growth, which exerts pressure on land. And also by the security crisis.

1.1.4 Research objectives

The general aim of this work is to show the impact of the security crisis and flooding on pepper production in the Diffa Urban Commune.

Specifically, this involves :

➢   Assessing the pepper production system in the Commune Urbaine de Diffa.

➢   Analyse the impact of the security crisis and flooding on pepper production.

➢   Identify producers' endogenous strategies for coping with security constraints and flooding.

1.1.5 Research hypotheses

In order to answer the questions posed in the problem, the following hypotheses are formulated:

➢   The pepper growing system produces good yields.

➢   The security crisis and floods have led to a drop in pepper production.

➢   Producers have adopted strategies to deal with security constraints and flooding.

---

[8] http://Www.Femininbio.com/alimentation/conseils-et-astuces/vertus-bienfaits-du pepper 54209, consulted on 4/02/2021 at 18:27.

1.1.6 Definition of concepts

The definition of key concepts is of paramount importance for any study or research in the social sciences. To make this document easier to understand, a number of concepts have been defined.

➤ **Impacts**

KOUAKOU A (2010) defines it as the tangible and intangible, direct and indirect, positive and negative effects that an incident, change, problem, movement or action has, or could have on a phenomenon. In the context of this work, impact refers to the positive and/or negative consequences of the *Boko Haram* security crisis on pepper production.

➤ **Security crisis**

Simply as "a lack of security, anxiety caused by the possibility of danger", it is in reality only the end result of threats formed by criminal movements and entities seeking to put an end to security, more or less global, and to impose their visions, with the aim of prospering better in time and space (ADAMOU M, 2019).

➤ **Growing system**

This is the use of an agricultural area by a group of plots with a homogeneous spatio-temporal organisation resulting in the same type of crop succession (JOUVE, 1984).

➤ **Peppers**

Peppers belong to the Capsicum genus of the *Solanaceaes* family, in the *Aristidae* subclass of the group of evolved dicotyledons characterised by gamopetalia (fused petals). *Solanaceae* belong to the order *Polemonic*, with a herbaceous habit and a superect ovary (GUIGNARD 1996).

➤ **Production**

In a common sense, production refers to the economic activity of creating goods and services. Production is first and foremost the result of human labour (JEAN. L, 2009).

➤ **Flooding**

Flooding is the submergence, whether rapid or slow, of an area that may be inhabited; it corresponds to the overflow of water (BOUBCHIR, 2007).

➤ **Adaptation strategies**

Adaptation strategies involve a dynamic of change or the adoption of new production techniques (ALPHA G, 2010). Adaptation can be spontaneous or planned, in response to or in anticipation of change.

1.2   Methodology framework

The methodology adopted for this study is broken down into several stages: documentary research, direct observation in the field, data collection, processing and analysis of the data collected.

1.2.1 Documentary research

Documentary research is a very important stage in any research project. Its purpose is to take stock of previous studies carried out in the field of this study. It enables us to make an inventory of the literature related to the subject. It also enables us to consult the works available in libraries. This documentary search is initially focused on general works relating to security problems in the Sahel and flooding. It also covers the various publications on peppers. These include theses, reports and dissertations. Internet sites were consulted, in particular the IRD website[9] using Google as a search engine.

This stage of the work provided important information on the subject, helping to define its various contours.

1.2.2 Site work

After researching the literature, surveys and interviews were used to gather the information needed to produce this report.

➢ **Direct observation of the field**

It consisted of direct field visits. During this phase, informal talks were held with producers. It also provided an opportunity to make contact with the administrative and customary authorities, as well as the government's technical services. In addition, this phase provided an opportunity to observe the biophysical characteristics of the study area.

➢ **Field data collection**

Data collection is a crucial part of the research process. This phase took place through a series of field trips. During this stage, quantitative and qualitative surveys were carried out to gather the necessary information. For the first surveys, a questionnaire was drawn up and administered to producers. For the latter, interview guides were drawn up. This made it possible to characterise the production systems, analyse the impact of the security crisis and flooding on pepper growing and identify the coping strategies adopted to deal with the insecurity and increase production. This stage was carried out in February and March 2022.

---

[9] IRD publications scoreboard https:// www.documentation.ird.fr

1.2.3 Sampling

Four (4) sites in the Commune Urbaine de Diffa were selected for this study, namely the CBLT, Lada Koulou, Bagara, and Tourban guida sites. The survey sites were selected on the basis of a number of criteria. These included their proximity to the Komadougou River, which is the pepper production area, the large number of producers at these sites, and the accessibility of the sites given the insecurity linked to Boko Haram. The target population is made up of pepper growers.

The surveys were based on a sample of 110 producers in the Commune Urbaine de Diffa, distributed as follows: (55) producers for the CBLT site; (22) producers for the Lada Koulou site; (22) producers for the Bagara site and (11) producers for the Tourban guida site. Clearly, this is simply simple random sampling, as the individuals making up this population were not grouped together before the draw, so the samples were chosen at random. The use of simple random sampling is explained by the difficulties associated with insecurity. From a total of 478 producers spread over the four sites selected in the Commune Urbaine de Diffa (Lada Koulou, CBLT, Bagara and Tourban guida), a sample of 23% was considered. This sample is representative given the security situation in the area and our very limited resources.

Table 1 Breakdown of producers surveyed by site

Sites	Number of producers	Number of producers surveyed	Number of producers surveyed (%)
CBLT	**239**	55	50%
Bagara	**96**	22	20%
Lada koulou	**96**	22	20%
Tourban guida	**47**	11	10%
Total	**478**	**110**	**100%**

Source: data from March 2022

Table 1 shows the distribution of respondents by site. The choice of 55 growers at CBLT is justified by the large number of pepper growers. In fact, the site is used by the inhabitants of the village and also those of the town of Diffa. However, at Bagara and Lada Koulou, pepper production remains marginal, so the number of growers is limited at these sites. Lastly, at Tourban Guida, the security situation is still unstable, making it impossible to survey several growers.

➢ **Quantitative survey (questionnaires)**

This approach was based on a questionnaire survey of pepper growers. The surveys were conducted over a one-month period from 25/02/22 to 29/03/22. It should also be noted that the questionnaires were only administered in the mornings between 8am and 2pm, given the worrying insecurity situation in the area. The information gathered relates to the pepper-growing system, the impact of the security crisis and flooding on pepper production, and local coping strategies. The questionnaire is divided into seven (7) sections:

- Socio-economic characteristics of respondents ;
- Land tenure for producers ;
- The security crisis and pepper production;
- Performance before and during the crisis ;
- Peppers and production methods ;
- Peppers and storage ;
- Local adaptation strategies

In short, this phase enabled us to gather the information we needed to carry out this work.

➢ **Qualitative survey (interviews)**

In addition to the questionnaire surveys, semi-structured interviews were conducted using an interview guide. A total of 4 interviews were conducted. These were chosen on the basis of the information they were likely to provide on the subject under study. Two village chiefs, a director of the regional chamber of agriculture and the president of the pepper traders were involved in this phase. The main topics discussed were land tenure, pepper growing, *Boko Haram* insecurity and the impact of the crisis and flooding on the pepper industry. This investigation was carried out using an interview guide

1.2.4 Data processing and analysis

The information collected was processed and analysed using Sphinx plus2-v5 software. This enabled the data collected through the quantitative surveys to be analysed. The data was then transferred to Excel to produce the graphs. The maps were designed using ARCGIS 10.3. Finally, Word was used to draft the present work.

1.2.5 Difficulties encountered

This theme deals with the *Boko Haram* security crisis in general and terrorism in particular in the Commune Urbaine of Diffa. It therefore naturally touches on defence and security issues. It is clear that in this context of insecurity, we are faced with a certain limitation, as it is difficult to distinguish between combatants and non-combatants. The problems of access to certain areas due to insecurity and the state of emergency mean that some players are inaccessible, but they are also limited in the information they can give out for fear of reprisals. In addition, there are difficulties in travelling and in choosing which producers to survey, given the state of emergency. And finally, the veracity of the data collected, because the intervention of several regional and international organisations necessarily influences the quality of the information collected.

Partial conclusion

This chapter provides a detailed presentation of the theoretical and methodological framework by setting out the problem, followed by a review of the literature. In addition, the hypotheses and objectives were presented, as well as the methodological tools used. In addition, the issue of data processing and analysis was addressed, and finally the difficulties encountered were presented. Following the theoretical and methodological framework, the study area is presented in the next chapter.

CHAPTER II: GENERAL PRESENTATION OF THE STUDY AREA

Introduction

In this chapter, the location of the study area, the physical environment, in particular the relief, climate, soils, flora and fauna and water resources, are presented, followed by demographic aspects and, finally, the socio-economic aspects of the study area.

1.3   Location of the study area

The Commune Urbaine of Diffa is located in the extreme south-east of Niger on the Route Nationale N°1, 1360 km from Niamey. It extends over a radius of 20km on either side of the urban centre, with an estimated surface area of 229km^2 . The Urban Commune of Diffa is bordered to the east and north by the rural commune of GUESKEROU and to the west by that of CHETIMARI. To the south, it borders the Federal Republic of Nigeria for more than 30 km, marked by the Komadougou Yobé PDC, (2014). Figure 1 shows the location of the commune and the survey sites.

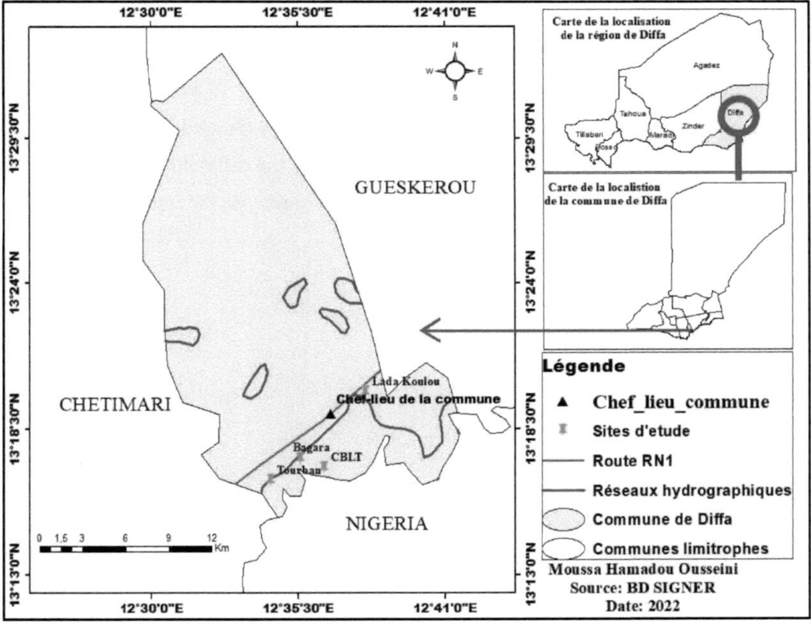

Figure 1 Location map of the Urban Commune of Diffa

1.4 Biophysical setting of the study area

This section presents the biophysical characteristics of the study area. These include relief, climate, soils, flora and fauna and hydrography.

1.4.1 Relief

The relief of the Diffa Region is shaped and characterised by lacustrine and alluvial influences to the south and aeolian influences to the north. It is made up of Tal, Manga and Kadzel sand dunes and Mandaran basins. There are no abrupt variations in topography except on the outskirts of the Agadem massif. The relief is made up of plains and plateaux, with altitudes ranging from 275m on Lake Chad to 550m on the Agadem massif. Apart from the Djajiri granite point to the west and the Agadem Cretaceous massif to the north, the outcrops are exclusively sandy-loam quaternary deposits, some of which are clay PDR (2020).

1.4.2 Climate

The Commune Urbaine of Diffa has a Sahelian climate characterised by a long dry season from October to June, divided into three (3) periods:

The first, October-February, has low temperatures with absolute minimums of between 6° and 25°c, favourable for off-season crops; the second, March-June, has high temperatures of between 33° and 46°c; and a short rainy season, July-September PDC, (2014).

At local level, apart from the officially recognised seasons, four other seasons have been identified during the year in the *Kanouri* language:

➢ The *Biyila* harvest takes place from September to October, and serves as a transition between the winter and the cold season.

➢ *Binoum* cold season from November to February;

➢ Hot season *Bé* from March to June ;

➢ *Nanguiri* rainy season from July to September.

The rainy season lasts a short three months at most, generally from July to September, with average rainfall of between 200 and 300mm per year PDC, (2014). It is also often marked by random rainfall, which is a source of drought and the gradual advance of the desert towards the south.

There are two types of wind: the harmattan, whose speed varies from 2 to 5 m/s at 10 m above the ground, blows from north-east to south-west for more than 5 months, from November to April. The monsoon blows from south-west to north-east from June to October. The average monthly wind speed varies between 1.3 m/s, observed in September, and 2.2 m/s, observed in July. For the local population, the change in wind direction heralds the arrival or end of the rainy season, which determines the bulk of agro-pastoral activities.

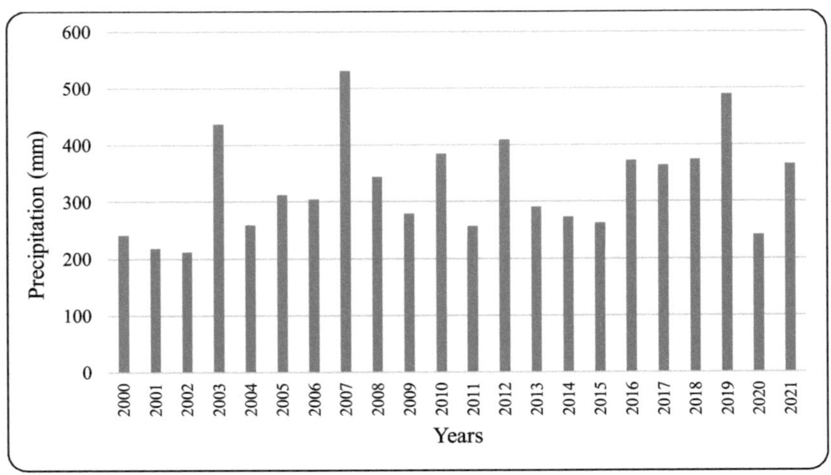

Figure 2 Rainfall trends at the Diffa station (DRM/D, 2022)

Source: (DRM/D, 2022)

Figure 2 shows cumulative rainfall over the last 22 years. Over the last 11 years, the average rainfall has been 336.23 mm, indicating an upward trend, with flooding in some parts of the Diffa Urban Commune.

1.4.3   Floors

The nature of the soil in the Diffa Region depends on the agro-ecological zones. MHE/LCD, (2004).

Three types of soil are found in the study area:

➤ Soils with partial surface hydromorphy: These are hydromorphic soils with pseudogley in association with other soils, particularly brown-red soils and vertisols. They are of average fertility, but their texture is unfavourable to internal drainage, their alkalinity is high and they contain little organic matter. This limits their suitability for cultivation over large areas. They are located along the Komadougou river and close to its banks; their potential use is for irrigated farming;

➤ Sandy clay soils, north of the RN1. These soils are complex brown and red, with average fertility and generally low organic stock. Dune crops such as millet, sorghum and cowpeas are grown here;

➤ Clay-loam soils along the meanders and also around the ponds. These soils are generally of good quality as they are made up of various types of alluvium. Irrigated pepper, rice, wheat, barley, onion and tomato crops are grown here PDC, (2014).

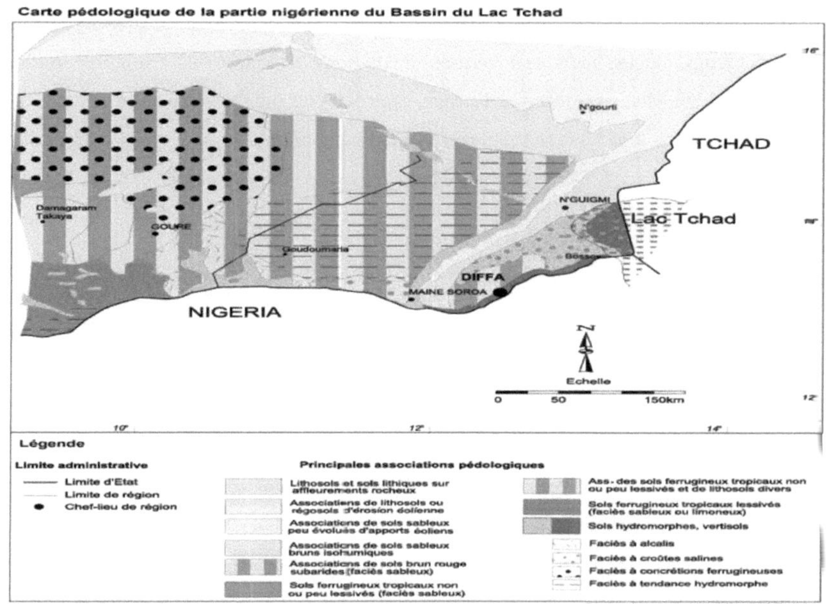

Figure 3 Soilology of the Niger part of the Lake Chad basin

Source: National Atlas of Niger DADT-2002

1.4.4   Flora and fauna

In the Commune Urbaine of Diffa, the vegetation is divided into two settlement zones from north to south. The vegetation is divided into two distinct categories shaped by the soil and climatic conditions:

➢   In the northern part of the town, the tree layer is dominated by urban plantations, notably *Azadirachta indica*, followed by natural stands of *Acacia radiana, Faidherbia albida Bauhinia rufescens, Balanites aegyptiaca, Acacia senegalensis,* etc.

➢   The southern strip around the Komadougou Yobé, the string of ponds and the banks in hydromorphic sandy-clay and silt-clay soils, is dominated by large trees such as *Tamaridus indica, Diospyros mespiliformis* and a few alkaline species such as *Hyphaene thebaica* as the tree layer. The shrub layer is sparsely represented in the area. Herbaceous plants are mainly aquatic forage species, which provide most of the grazing during the dry season.

➢   **Wildlife resources**

Two distinct wildlife zones have been identified. These zones are based on soil composition and climatic phenomena. They include :

➤ The Komadougou area is home to primate monkeys and birds such as crowned cranes, egrets, herons, ducks, teals and various small birds and migrants from the southern hemisphere, the Mediterranean and Europe, especially during the wet season;

➤ Area north of the RN1, where reptiles, rodents and countless insects, such as butterflies, bees, etc. can be found. Domestic fauna includes cattle, sheep, goats, camels and poultry DRESU/DD, (2018).

1.4.5 Water resources

After Niamey, Dosso and Tillabéry, the Diffa region has the best supply of surface water. In fact, an average of 500 million m^3 of water passes through the Bagara reference station on the Komadougou Yobé, to which must be added the waters of Lake Chad and numerous strings of ponds COMMISSION DEVELOPPEMENT RURAL, (2006).

The hydrographic network of the Urban Commune of Diffa is dominated by the Komadougou Yobé and its meanders, which form a complex of permanent and semi-permanent pools.

➤ Komadougou Yobé: is a semi-permanent watercourse that provides most of the surface water resources with Lake Chad in the Diffa Region. This river has its source in Nigeria and then flows into Lake Chad. The area around the river is characterised by meanders and hydromorphic soils favourable to off-season crops (ABDOU A et *al.*, 2005). In addition, in the downstream area of the basin, where the river forms the border between Niger and Nigeria, a significant amount of market gardening, particularly of sweet peppers, has developed (HADIZA K, 2014). Its flow is temporary, beginning with the rainy season in the first half of July, peaking in November or December and becoming nil a few weeks later. The river flows for an average of 7 months a year, with gradual flooding from June to peak flow around November, followed by rapid recession until it dries up around February-March (MOUSSA I, 2014). The Komadougou carries an average of 500 million m3 of water per year. It is a degressive river that loses much of its water through infiltration, spreading and evaporation, mainly in its Niger course (CRAT, 2018).

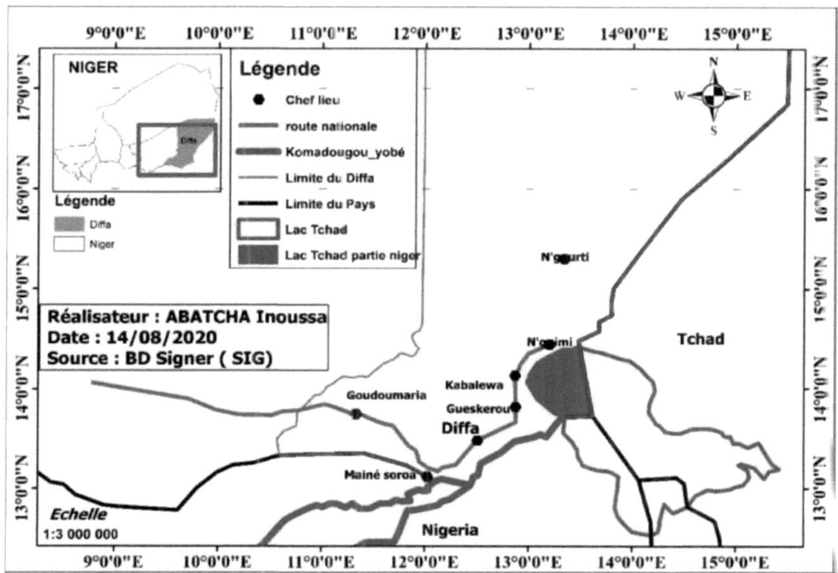

Figure 4 Location of Komadougou Yobé

Source: Signer database (GIS, 2020)

➢ Ponds: these are seasonal and collect rainwater or water from the Komadougou Yobé. They are mainly located near the Komadougou, where they are the old meanders. They can be classified according to their water regime. There are permanent pools and semi-permanent pools.

➢ Underground water resources

In the Urban Commune of Diffa, groundwater can be distinguished according to the aquifers shaped by geological phenomena. To this end, four aquifers have been identified in terms of depth and lithological variation: DRHA, (2018)

• **The Manga aquifer**, 10 to 50 m deep and 50 to 80 m thick. It is the most accessible for the construction of traditional, village and pastoral wells and a few boreholes. The most common problem encountered is the lack of improvement in catchment techniques, which are better suited to the nature of the aquifer;

• **Pliocene aquifer,** deeper than the first, 300 m on average and 8 to 85 m thick. It is made up of fine to coarse sands separated by clays. It is the main source of drinking water for the town of Diffa, as it is very deep and free from pollution;

- **This deep aquifer** is one of the sources that has never been exploited due to a lack of technical and logistical resources specifically dedicated to its nature;

- **Alluvial phreatic aquifer of the Komadougou Yobé or superficial aquifer,** which is close to the major bed of the Komadougou. It is recharged by surface run-off from the Komadougou and is the main source of water for the manga.

1.5   Demographic aspects

This section presents the demographic characteristics and socio-economic activities of the Urban Commune of Diffa. These include agriculture, livestock farming, fishing, trade and handicrafts. These five areas of production constitute the main socio-economic activities of the population of the Urban Commune of Diffa.

> ➢ **Demographic characteristics**

In 2012, the Diffa Region was home to the least populated sedentary population in the country, estimated at 939,988 inhabitants, representing a density of 5.9 inhabitants/km2 in 2022. This population is unevenly distributed. The Urban Commune of Diffa, comprising 6 districts, accounts for 68.87% of the population. The most populous district is Sabon-carré, followed by Diffa-Koura, Festival, Diffa-administratif and Château. The least populated district is Diffa-Afounori. The 21 villages that we consider to be the rural part represent around 31.13% of the population. But unlike the urban area, women outnumber men here, with 50.48% and 49.52% respectively PDC, (2014). With a population growth rate of 4.7% and a total fertility rate of 6.4 children per woman DR-INS Diffa, (2018). The population is relatively young, in line with that of the country as a whole, with 51.2% under the age of 15 DR-INS Diffa, (2018). The largest age groups are 0-6, 7-12 and 13-18. Together, these three age groups make up 51.2% of the total population. The smallest age group is 49-55, with 3% of the total.

Table 2 Some demographic indicators

Demographic indicators	Diffa	Niger
Population under 15 (%)	51,2	52,1
Population aged 65 and over (%)	3	2,6
Urbanisation rate (%)	14,8	19,8
Population growth rate (%)	4,7	3,9
Density (inhabitants / km2)	4,4	11,3

Total fertility rate (EDSN-MICS 2021)	6,4	7,4
Infant mortality rate (‰)	18	51
Infant mortality rate (‰)	41	127

Source: Database (NSI, 2018).

1.6   Socio-economic aspects

These five 05 areas of production constitute the main socio-economic activities of the population of the Diffa Urban Commune. These are agriculture, livestock farming, fishing, trade and handicrafts.

1.6.1   Agriculture

Agriculture is one of the main socio-economic activities of the people of Diffa. It mainly takes the form of cash crops or food crops. There are 2 cropping systems: rain-fed and irrigated.

➢   **Rainfed crops,**

The rainfed system is characterised by agricultural production based on cereals in association with legumes. Production is unstable, partly because of the vagaries of the weather, but also because of fluctuations in the area under cultivation and crop pests.

The main crops are millet, which is widespread in the south-eastern fringe of the region; in addition to millet, sorghum and maize are grown in the lowlands and in the Komadougou and Lake Chad valleys; cowpeas, groundnuts, sesame and sorrel are grown on dune soils.

Table 3 Production trends for the main rainfed crops from 2015 to 2021 in Diffa department.

Crops	2015	2016	2017	2018	2019	2020	2021
Mil	20681	20681	14548	15491	8502	9457	7808
Sorghum	3701	3701	3021	3975	1854	2360	2601
Cowpeas	9429	9429	5279	5743	3939	4700	5858
Peanut	2354	2400	1018	1608	1145	900	1433
Total	36165	36165	23866	26817	15440	17417	17700

Source: (DRA DIFFA, 2022)

Generally speaking, from 2015 to 2016 there was an increase in production, particularly for millet, sorghum and cowpea. Groundnut production, on the other hand, remained virtually constant from 2017 to 2019. However, production of millet and groundnuts has fallen due to rainfall and a significant reduction in acreage. Production of the main rainfed crops has fluctuated up and down.

➢ **Irrigated crops.**

The irrigated system is based on the region's enormous potential for irrigated and flood recession crops. Irrigated crops have therefore been introduced to make up for shortfalls during the winter season. The main crops are peppers, rice, onions, cabbage and potatoes. The region has an irrigable potential of around 183,000 ha in three sub-zones:

➢ **The "cuvettes" zone** covers an area of around 8,000 ha and is home to a variety of crops including cassava, wheat, sweet potatoes and sugar cane, as well as fruit trees such as lemon, papaya, guava, mango and, above all, date palms;

➢ **The Komadougou area,** covering some 75,000 ha, is characterised by the development of rice, pepper, onion, wheat and various vegetable crops;

➢ **The lake zone**, covering around 100,000 ha, is where maize and sorghum are grown. CRAT, (2018).

The main crop grown under irrigation, pepper production covers around 4,000 to 7,000 ha per year, with a production of around 11,000 tonnes and an overall gross income of 6 to 10 billion CFA francs. According to the Diffa CRA, in 2014 production was estimated at 11,143 tonnes of fresh peppers, or 655,460 bags of dried peppers, for a total income of 19,663,800,000 CFA francs. However, this sector has recently been threatened by climatic hazards, pest attacks and the threat of attacks by the terrorist group *Boko Haram* CRA, (2020).

Table 4 Areas, yields and production of the main irrigated crops from 2020 to 2021 in Diffa department

Speculation	Area in hectares		Yield in Kg/hectare		Production in tonnes	
	2020	2021	2020	2021	2020	2021
Peppers	2866,60	2885,75	14,11	23,20	40459,16	66949,40
Rice	1247,00	1375,60	1,750	5,75	2182,25	7909,70
Onion	358,46	395,60	1,95	27,85	697,92	1107,46
Cabbage	283,16	352,60	23,76	24,56	6727,88	8659,86

Source: (DRA DIFFA, 2022)

1.6.2 Breeding

Livestock farming is the dominant economic activity in the Diffa Region, with a highly diversified herd. It employs 95% of the population and contributes 55% annually to the region's gross domestic product. However, this activity faces various biophysical and anthropogenic constraints (ABDOULKADRI L, 2014). Livestock farming is practised by nomads but also by sedentary people who also practise agriculture. The region has a fairly large stock of cattle, sheep, goats, camels, horses and donkeys. Two types of livestock farming are practised in the Urban Commune of Diffa: Sedentary livestock farming is characterised by the movement of livestock back and forth from the town of Diffa to the grazing areas along the Komadougou Yobé, as well as to the various accessible watering points and veterinary treatment centres. They account for 64% of the livestock in Diffa department. Nomadic herding is practised by nomadic Fulani herders along the Komadougou Yobé during the rainy season and to a lesser extent in the rain-fed fields during the dry season. In turn, it accounts for 16.4% of the livestock in Diffa department. Transhumance takes place through a south-north pastoral movement, with areas of livestock concentration in the north during the rainy season and movements down to the south in search of water and pasture during the dry season MINISTERE DE L'EQUIPEMENT, (2020).

Table 5 Livestock trends in Diffa department from 2015 to 2021

Years	Cattle	Sheep	Goats	Camels	Equines	Asins	Total
2015	186 351	142 475	236 346	21 563	10 224	45 004	641 963
2016	197 532	147 461	245 800	21 844	10 326	45 904	668 867
2017	209 384	152 623	255 632	22 128	10 430	46 822	697 017
2018	221 947	157 964	265 857	22 415	10 534	47 758	726 476
2019	235 263	163 493	276 492	22 707	10 639	48 713	757 308
2020	339 599	169 022	287 127	22 999	10 744	49 668	879 159
2021	421 975	174 551	297 762	23 291	10 849	50 623	979 051

Source: (DRE Diffa, 2022)

1.6.3 Fishing

Fishing is one of the major activities in Diffa Region, and relies exclusively on natural resources. Fish production in Diffa Region comes mainly from Lake Chad and the Komadougou Yobé CRA river (2010). The fish species that inhabit the waters of the Diffa region are *Tilipiasp, Clarias griepinus* and *Heterotis niloticus*. Fish production in the Diffa region accounts for more than 70% of national production. It makes a significant contribution to GDP growth. Fish catches vary between 6,000 and 20,000 tonnes of fresh fish equivalent, depending on the extent of the lake in the national territory. Fishing is carried out by around 10,000 professional fishermen and a few agri-fishermen using longlines, gillnets, sledgehammers, hawks, stages and the "*Góra*" or dugout canoe. Lake Chad, the Komadougou Yobé and its systems of ponds make up most of the fishing potential of the Diffa Region CRAT, (2018). It should also be noted that the current insecurity linked to the *Boko Haram* security crisis has reorganised fishing territories and the fish industry. The current dynamics of fishing and the fish industry in the Diffa Region are the result of this recent history (HADIZA K, 2019).

1.6.4 Trade

Trade is organised around the central market, the fish market, the pepper market, the livestock market, the modern abattoir, a bus station and around the town centre. These markets sell products such as peppers, livestock, fish and some manufactured goods exported from Nigeria. Foreign trade is mainly with Nigeria and involves livestock, peppers, hides and skins, natron and transit products such as rice, tea, sugar, oil and cigarettes. Trade with Nigeria is undeveloped and informal. The closure of the border with Nigeria, with the advent of insecurity, has considerably slowed down trade PDC, (2014).

1.6.5 Crafts

The Diffa Region has a large number of craftspeople, estimated at 1,145, grouped into several trades, the majority of whom live in rural areas CRAT, (2018). Today, craft activities are mainly concentrated in the major centres of Diffa, Mainé and N'guigmi and practised in the traditional way. Craft activities in the region include Blacksmithing, basketry, shoemaking, leatherwork, pottery, butchery, sewing and carpentry. It should also be noted that handicrafts are mainly consumed locally, as sales through external markets such as neighbouring Nigeria and the tourism circuit are not very well explored.

Partial conclusion

This chapter provides a geographical overview of the study area and presents its main biophysical characteristics. The demographic and socio-economic characteristics of the area are also presented.

At the end of this chapter, the assets, potential and obstacles of the study area make it a potentially rich and favourable location for irrigated crops.

CHAPTER III: OVERVIEW OF PEPPER PRODUCTION SYSTEMS

Introduction

This chapter locates the pepper production area. It then presents a brief history of pepper growing in the Diffa region. Finally, the general characteristics of pepper growers and production systems are presented. These include the age of growers, their level of education, their access to land and the typology of growers according to the area under cultivation and cultivation practices.

1.7 Geographical location of the pepper production area

In the Diffa Region, pepper production is practically concentrated along the Komadougou Yobé Figure 5. More precisely, this production zone is located in a strip about 150 km long and 5 km wide between Bosso and Kanama, where it is the main cash crop and occupies 60 to 65% of irrigated land (PIERRE-FRANÇOIS P and SALIFOU K, 2005). Overall, the region has an estimated 265,000ha of exploitable land, including 182,000ha on the bed of Lake Chad, 75,000ha along the Komadougou Yobé and 8,000ha in the Maine Soroa PDR oasis basins (2020).

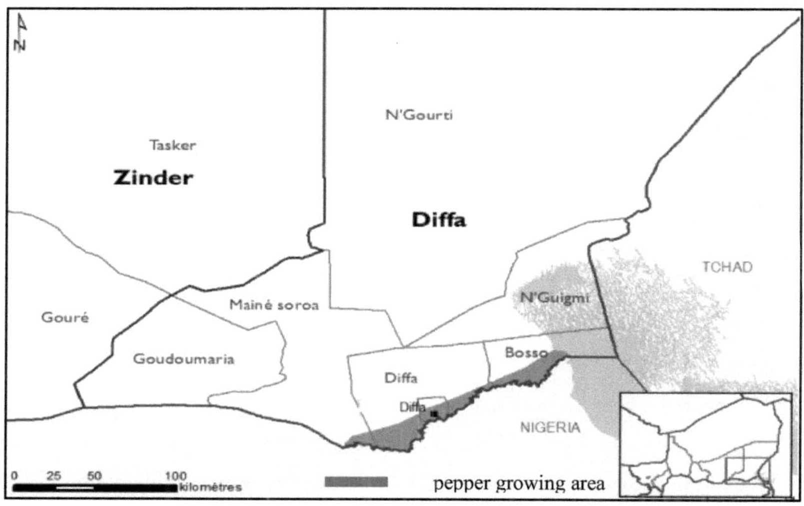

Figure 5 Location of the Diffa pepper-growing area

Source : FEWS NET

1.8 History

Irrigated pepper growing as a cash crop was introduced around the 1950s in Diffa department, but it only took off in the 1980s following repeated droughts and price rises, which made this speculation one of the sources of income for rural households in Manga (KORE and HAMIT, 1991 cited by OUSMAN Y, 2014). Indeed, the spread of small motor pumps adapted to individual farming led to a considerable increase in production and, in the 2000s, from this strip 150 km long by 5 km wide, nearly 10,000 tonnes of dried pepper were exported annually, representing a windfall of 7 to 8 billion CFA francs (€7 to 12 M) benefiting a farming population estimated at over 30,000 people (PIERRE-FRANÇOIS P and SALIFOU K, 2005 cited by LE COZ M, 2010).

1.9 Characterisation of producers

This section presents the characteristics of the producers surveyed in the study area. These include age, level of education, access to land and the typology of pepper growers.

1.9.1 Age of producers

In the study area, the age of pepper growers varies between 19 and 71 Figure 6.

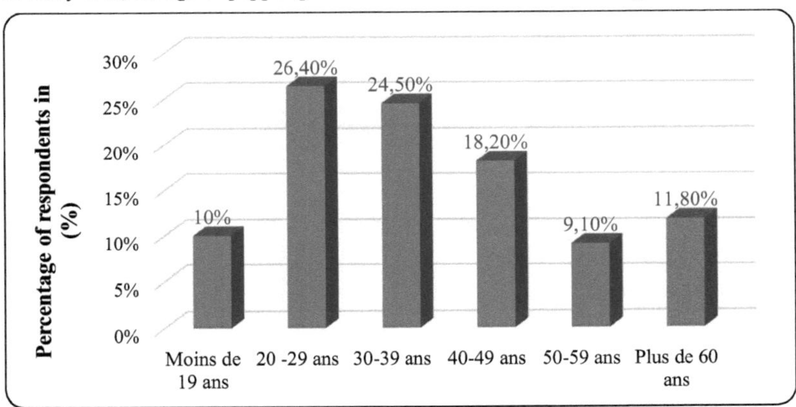

Figure 6 Breakdown of respondents by age group

Source: field data, 2022

Figure 6 shows the age ranges of pepper growers. These range from under 19 to over 60 years of age. Analysis shows that 50.9% of growers are aged between 20 and 39, with only 10% aged under 19. This shows that producers in the study area are young. However, the oldest producers represent only 11.80%.

1.9.2 Level of education of producers

According to the quantitative surveys, farmers have a very low level of education Figure 7. This low level of education has an impact on farm performance.

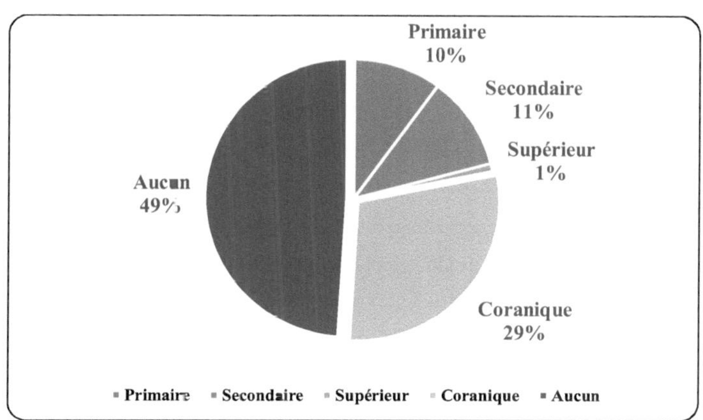

Figure 7 Distribution of respondents by level of education

Source: field data, 2022

Analysis of figure 7 reveals that 49% of farmers in the study area are illiterate, i.e. they can neither read nor write and have no formal education. A further 29% had attended Koranic school. On the other hand, 11% had secondary education, 10% had attended primary school and only 1% had higher education.

1.9.3 Ground access mode

According to the quantitative surveys, there are four 4 methods of acquiring land in the study area. These are inheritance, rental, purchase and loan. The extent of the latter varies Figure 8.

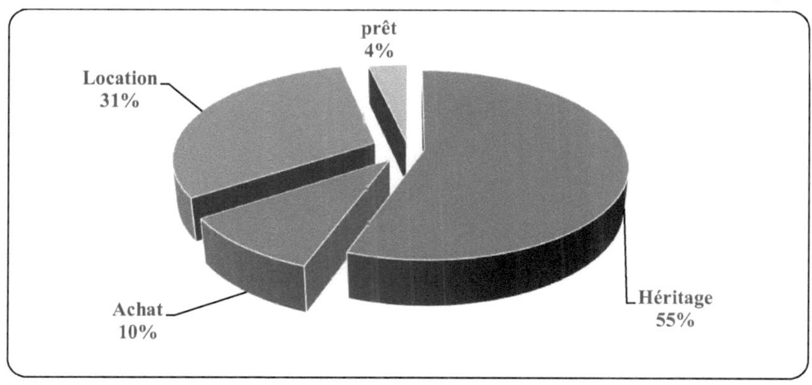

Figure 8 How farmers acquire land

Source: field data, 2022

Analysis of Figure 8 shows that inheritance is the main form of access to farmland, accounting for 55% of respondents. The second most common form of access is renting, which is the second most common form of land acquisition. Those who rent land represent 31% of the sample surveyed. The third mode of land acquisition is purchase. Farmers who bought their land accounted for 10% of those surveyed. And loans, which concern non-owner producers who develop borrowed farms, make up only 4% of the sample.

1.9.4 Typology of producers according to area

Pepper growers were classified according to the area under cultivation. Three categories of producers were identified: small producers, medium producers and large producers. Figure 9 shows that

➢ Producers with a holding of less than 0.99 ha are classified as small producers;

➢ And those whose farms cover between 1-1.99 ha are referred to as medium-sized producers;

➢ Finally, those with a surface area of more than 2 ha make up the large producers.

Figure 9 shows the status of growers according to the area farmed in the study area. The analysis shows that 44% of farmers are small-scale producers, 47% are medium-scale producers and only 9% are large-scale producers.

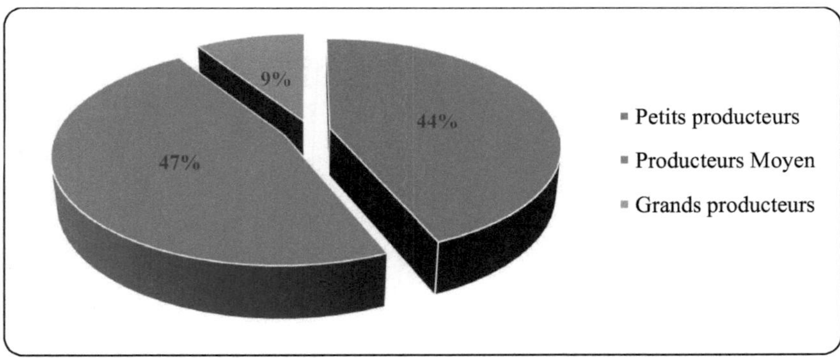

Figure 9 Distribution of farmers between the 3 categories defined by cultivated area

Source: field data, 2022

1.10 Cultivation practices

Cultivation practices are divided into seven 09 stages. This section successively presents the cultivation system, the work involved in preparing the plots and transplanting, and the irrigation system. Then the pepper varieties, the seeds, the cultivation calendar, the use of inputs and pesticides and finally the drying and conservation of the pepper.

1.10.1 Nursery

The nursery is used to raise the plants that will later be transplanted. Setting up nurseries is the first cultivation operation carried out by pepper growers. The nursery has a well-developed, fenced-in area where the pepper plants are sown and cared for. Each grower has his own nursery in his house or in his field (photo 1). In general, growers set up the nursery in May-June before the first rain. The nursery is watered twice a week.

Photo 1 Pepper nursery

Source: field data, 2022

1.10.2 Preparing plots and transplanting

Preparation of the plots begins in July and continues until August. It is carried out by hand by the growers or with the help of animal traction. Work is carried out using rudimentary tools such as hoes and dabas. When the soil is too clayey, sand is added to the plots to reduce the clay content. The plots are then amended with organic manure. The work of preparing the plots is done when the Komadougou Yobé arrives. Once this initial work has been completed, the farmers start transplanting.

- Ploughing

Ploughing is part of plot preparation. It is carried out using archaic equipment such as the hoe or the daba for some farmers, and the plough for wealthier farmers.

1.10.3 Transplanting

Transplanting consists of removing small plants from a nursery and transplanting them to the production plot. Transplanting peppers is done manually. This operation begins in July and can continue until the end of August or even longer. Everything depends on the water supply from the Komadougou and the rain.

1.10.4 Weeding

Weeding is a task that takes place after transplanting and consists of eliminating weeds that prevent the plant from developing. The frequency of weeding for pepper production is 3 to 4 times a season, or even more if necessary.

1.10.5 Irrigation system

According to the interviews, water is not a problem for growers. The Komadougou river is the main source of water for plants on production sites. The pepper plots are irrigated from the riverbed or from backwaters and ponds fed by the Komadougou. Most production is carried out by growers operating individual plots using small motor pumps imported from Nigeria.

➢ **Water pumps**

Motor-driven pumps are used to pump water from boreholes or water points (photo 2). They are used by farmers whose fields are located many kilometres from the Komadougou river. The motor-driven pumps feed a mini distribution network through a system of earthen gravity channels.

Photo 2 Motor-driven pump for irrigation from a water point

Source: field data, 2022

➢ **Drilling**

Based on the qualitative surveys, manual drilling is a system that gives farmers access to irrigation water at low cost. Manual drilling is carried out using hand augers and pipes in unconsolidated soil up to 12 metres deep (Photo 3). Water is generally distributed by the forge via a network of earthen canals, roughly furrowed to bring the necessary quantity of water to the plots.

Photo 3 Irrigation well for peppers

Source: field data, 2022

➢ **Gravity systems**

Growers use the gravity-fed system because their fields are directly adjacent to the Komadougou. The proximity of the Komadougou river means that growers use the gravity system to irrigate their pepper plots. In the gravity-fed system, water is distributed using a network of earthen canals in the form of furrows, which use gravity to bring the necessary quantity of water to the plots (photo 4).

Photo 4 Irrigation of peppers by gravity

Source: field data, 2022

1.10.6 Pepper varieties

Based on direct observations on the sites, the two existing local cultivars appear to be well adapted to the environment. Their names in Kanouri illustrate the shape, which translates as "sheep's horn" for the cultivar found mainly in the west, and "donkey's dung" for the cultivar more commonly grown in the east. CRA, (2010). Producers use the two local varieties in the Commune Urbaine de Diffa, *Kangadi N'Glaro* corne de mouton photo 5 A and *Mouri Koro* crottin d'âne photo 5 B, of which over 90% of respondents produce *Kangadi N'Glaro* according to quantitative surveys.

Photo 5 Photo **A** *Kangadi N'Glaro* (sheep horn) Photo **B** *Mouri Koro* (donkey dung)

Source: field data, 2022

1.10.7 Seed

In all the study sites, seed is not a problem for pepper growers. Farmers produce their own seeds. The seeds used by the growers are mostly local "crottin de l'âne" and "corne de mouton" varieties. They come from previous seasons' production. Producers keep their seeds in order to use them for several years. According to the quantitative surveys, 96.36% of respondents use local seed saved by producers. Only 3.64% of producers use seeds bought on the market.

1.10.8 Cultivation calendar

Table 6 summarises the different phases of the pepper cultivation calendar in the Commune Urbaine de Diffa. Field preparation begins in early July, coinciding with the arrival of the Komadougou River and the start of the rainy season. The harvest periods are December-January-February and March respectively. The pepper growing calendar extends almost all year round due to its relatively long vegetative cycle. However, it is difficult to establish a crop calendar for peppers due to changes in the rainy season and the arrival of the Komadougou River in recent years.

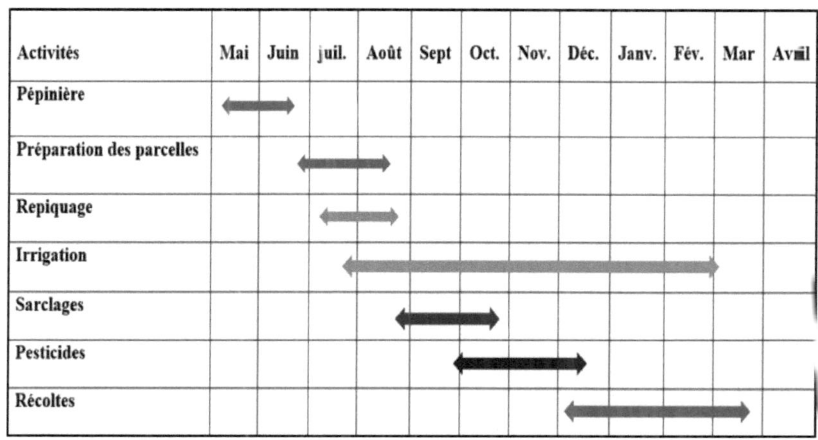

Table 6 Cropping calendar

Activités	Mai	Juin	juil.	Août	Sept	Oct.	Nov.	Déc.	Janv.	Fév.	Mar	Avril
Pépinière	←	→										
Préparation des parcelles		←	→									
Repiquage			←	→								
Irrigation				←							→	
Sarclages					←	→						
Pesticides						←	→					
Récoltes									←		→	

<u>**Source:**</u> field data, 2022

1.10.9 Use of inputs

According to the results, fertiliser is used extensively by farmers. The results of the Figure 10 surveys show that more than 68% of the sample use mineral fertiliser supplied on local markets, in particular the Diffa pepper counter and the Diffa central market. A further 23% of the sample use both mineral and organic fertilisers. And only a tiny 9% of producers use animal manure. Animal manure also helps to maintain the soil's natural fertility.

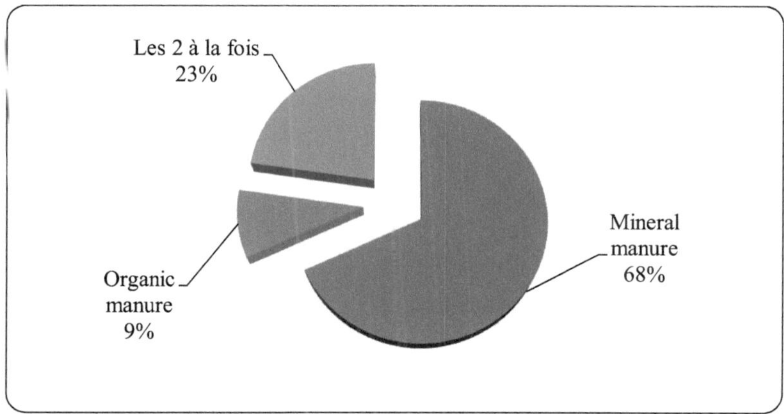

Figure 10 Use of agricultural inputs

<u>**Source:**</u> field data, 2022

1.10.10   Pesticide

Pesticides are chemical substances used to combat harmful organisms and pests. These substances are capable of killing or repelling pests and insects. These products come in a wide variety of sizes and forms (Photo 6).

These are substances capable of either killing or repelling pests, insects, mites and nematodes. The use of pesticides is crucial to pepper production, as it determines yield. Most of the plant protection products used come from Nigeria, and 100% of those surveyed said they had used pesticides. Producers buy their produce on local markets, in particular the Diffa pepper counter and the Diffa central market, from resellers.

Photo 6 Some samples of pesticides used for treatment.

Source: field data, 2022

---

[10] Regional Plant Protection Services (SRPV).

Table 7 Pesticides supplied by the State and pesticides sold on the CUD market

Pesticides fournis par l'Etat		Pesticides en vente sur le marché de la CUD		
Nom commercial	Matière active	Nom Commercial	Matière active	Autorisé/Non
Pyrical 480	Chlorpyriphos Ethyl	Assial	....	Non autorisé
Pyrical 240	Chlorpyriphos Ethyl	Farin	....	Non autorisé
Zalang	Lamda-cyalothrine	Megaban	....	Non autorisé
Fénical	Fenitrothion	Dursban	....	Non autorisé
Deltacal	Deltamethrine	Lara force	Lamda-cyalothrine	Non autorisé
Pacha	Lamdacyhalothine – Acétamipride	Marshal	....	Non autorisé
Capt 88	Acétamipride + Cypermethrine	Lambdashi	Lamda-cyalothrine	Non autorisé
Conquest	Acétamipride + Cypermethrine	Lymo	Lamda-cyalothrine	Non autorisé
<u>Source</u>: SRPV[10]/Diffa		Sunhalothrin	....	Non autorisé
		Perfect Killer	Chlorpyriphos	Non autorisé
		Rocket	....	Non autorisé
		<u>Source</u>: donnée terrain		

Source: field data, 2022

As part of its support policy for growers, the Niger government buys approved products. In fact, these products are used by the plant protection services for treatments by plane or lorry, as well as by the phytosanitary brigadiers (table 7). The majority of pesticides on the market are not registered. However, some pesticides such as Lara force, Lambdashi, Lymo and Perfect Killer contain a registered active ingredient and are not authorised by the government, but are widely used by growers.

1.10.11   Drying and storing peppers

According to the interviewees, the ripe fruit is harvested by hand and placed in an open area. The fruit is then spread out on a drying area, turned over by hand or with the feet to expose it to the sun, and dried on the ground in the sun or shade for 15 to 25 days by the growers. Drying is done in the sun at the edge of the plot Photo 7 B. For storage, the peppers are packed in jute bags covered in plastic Photo 7 A to preserve their quality and prevent termite attack. The bags are stacked in the shop. Peppers can be stored for several years without losing their quality. The bag weighs around 17kg CRA, (2016) Peppers are easy to dry and store. Properly dried, it can be stored for several years without significant loss. Moreover, storing peppers poses no problems. When dry, they are not attacked by insects.

Photo 7 Photo A: Pepper storage Photo B: Pepper drying

Source: field data, 2022

Partial conclusion

In short, the characteristics of pepper growers revealed that the majority are adults. The pepper-growing system in the Commune Urbaine de Diffa is still traditional, with the use of local seeds and rudimentary agricultural equipment. The pepper plots are irrigated from the riverbed or from backwaters and ponds fed by the Komadougou. However, the drainage system has improved considerably with the introduction of motor-driven pumps and then boreholes.

CHAPTER IV : THE IMPACT OF THE SECURITY CRISIS AND FLOODING ON PEPPER PRODUCTION AND LOCAL ADAPTATION STRATEGIES

Introduction

The aim of this chapter is to analyse the impact that the conflict caused by *Boko Haram* and the military operations to combat it have had on pepper production, as well as the effects of recurrent flooding and the local coping strategies used by producers in the study area.

1.11 Analysis of the *Boko Haram* security crisis on pepper production

The security crisis caused by *Boko Haram* has had a number of negative effects on the population, particularly on their livelihoods. With the intrusions of *Boko Haram* into Niger the security situation has changed radically for producers in the Diffa region, and especially in the Komadougou valley. Pepper growers have been severely affected by the security crisis.

1.11.1 Situation of producers

According to the field survey, the data show that a significant proportion of producers have ceased their activities because of the security crisis. These were producers affected by the security crisis, farmers affected by the floods and those who continued production (Figure 11).

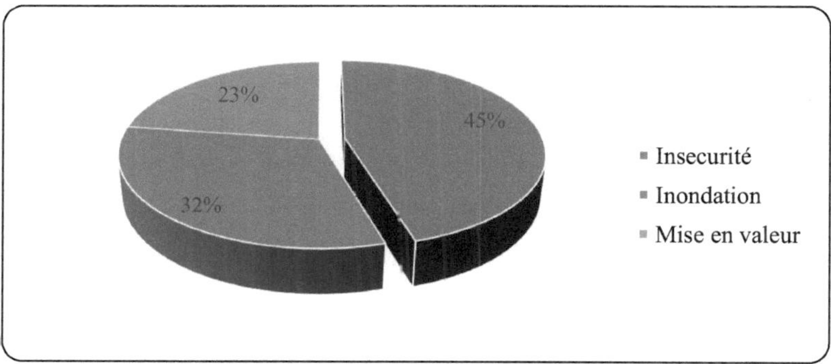

Figure 11 Situation of producers in the study area

Source: Field data, 2022

Figure 11 shows the situation of farmers in the study area. Analysis of the figure shows that 45% of the farmers surveyed stopped growing peppers as a result of the *Boko Haram* security crisis, compared with 23% who continue to cultivate their land despite the complicated security situation. The farmers surveyed mentioned that they face security risks on a daily basis. In addition, 32% of those surveyed have given up production since 2019 because of the recurrent flooding that has occurred every year since then.

1.11.2 Impact of the *Boko Haram* crisis on cultivated areas

The *Boko Haram* security crisis has led to a significant reduction in the area under pepper cultivation in the study area. The Boko Haram threat has considerably affected the area under pepper cultivation (Figure 12).

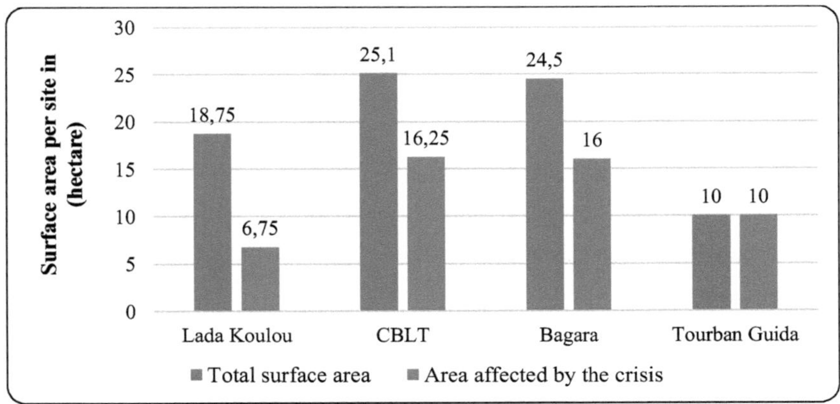

Figure 12 Area affected by the security crisis

Source: field data, 2022

Faced with this situation of insecurity, Figure 12 shows that 49.6 hectares, or 46.29% of the total area, have been abandoned by their owners. According to interviews, several farmers have been victims of reprisals by *Boko Haram* fighters. In addition, recurrent attacks by the terrorist group on military positions have prompted some growers to abandon their production land. In addition, the activities of the Islamic sect on producers, such as the collection of taxes, have forced some producers to cease their activities.

1.11.3 Impact of the *Boko Haram* crisis on pepper production

In the study area, the security crisis has considerably reduced pepper yields (Figure 13). Indeed, the units of measurement most used by growers to quantify pepper production are the sack corresponding to 17 kg of dried peppers.

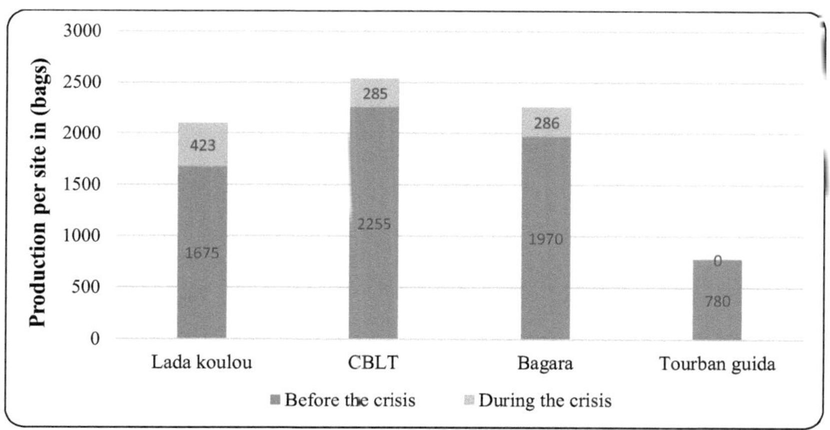

Figure 13 Production by site before and during the security crisis

Source: field data, 2022

Figure 13 shows that the cumulative production of the study sites before the security crisis was 6680 bags of dried peppers. This quantity varies between the CBLT site with 2255 bags, then 1970 bags for the Bagara site, 1675 bags for the Lada Koulou site and finally 780 bags for the Tourban guida site. With the crisis, cumulative production at the study sites fell to 994 bags (a drop of 85.11%), with one site, Tourban Guida, having 0 bags due to a break in production. This considerable drop in production clearly shows the influence of *Boko Haram* on this activity.

1.12   Impact of flooding on pepper production

The floods have had a severe impact on irrigated crop production, particularly sweet peppers in the Commune Urbaine de Diffa. All 4 study sites were affected by the flooding, but the worst affected sites were CBLT (71%) and Lada Koulou (14%).

These floods caused various types of damage to production and arable land. Table 8 shows the results of flooding in the Commune Urbaine de Diffa in 2019, according to the Diffa Departmental Directorate of Agriculture. This resulted in significant material damage, loss of life for 02 people and 580.21 hectares of rice and pepper crops swallowed up.

Table 8 Flood damage in the Commune Urbaine of Diffa in 2019

Balanc e sheet CUD	Household s affected	People affecte d	People Decease d	Collapse d houses	Market garden crops destroye d (ha)	Rice and pepper crops destroye d (ha)	Damage d wells (number
	1 075	18 690	2	2877	170,72	580,21	2

Source: Diffa Departmental Directorate of Agriculture 2019

1.12.1 Natural causes

These include meteorological and climatic events such as exceptional rainfall and flooding on the Komadougou Yobé.

1.12.2 Precipitation

Rainfall is one of the main natural causes of flooding in the Commune Urbaine de Diffa. The Commune is located in the Sahelian zone, which is characterised by a long dry season and a short rainy season, the latter running from July to September with an average rainfall of between 200 and 300mm per year.

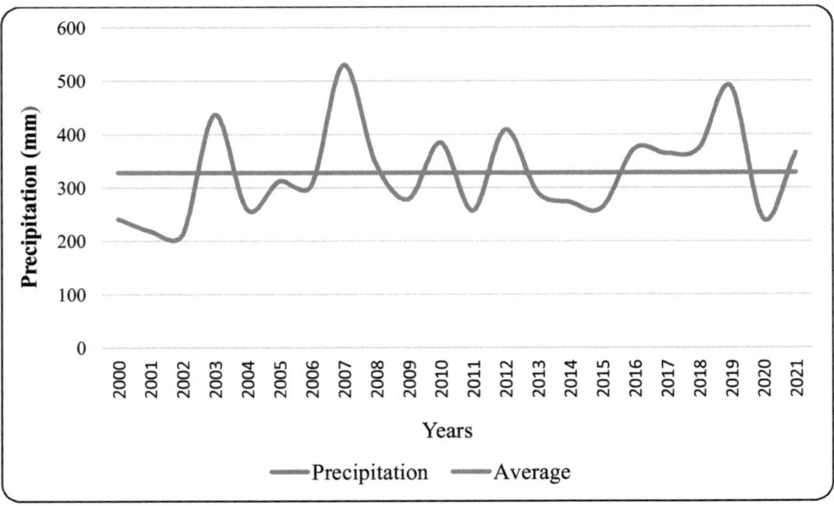

Figure 14 Cumulative annual rainfall over the last 22 years

Source: DRM/D 2022

Figure 14 shows cumulative rainfall from 2000 to 2021. The average rainfall over the last 22 years is 328.18 mm. These years are characterised by abundant rainfall, often causing flooding. In fact, 2016, 2017, 2018, 2019 and 2021 are the years with the highest rainfall. The rainfall recorded was 372.3, 364.1, 373.7, 489.3 and 365.3 mm respectively. This is above the average of 328.18mm. This led to recurrent flooding of pepper production.

1.12.3 Flooding of the Komadougou Yobé

All the study sites are located along the Komadougou Yobé river, which is the main source of surface water. Exceptional high water levels in this river cause flooding along its course. According to the Direction Régionale de l'Hydraulique, the data collected over the last three decades shows a continuous increase in the volume of water in the Komadougou river. This causes recurrent flooding in the commune. The data shows that the areas located in the Komadougou valley and the Diffa commune are constantly on alert, with water levels exceeding 4.49 m, the alert level. This puts the area at constant risk of flooding.

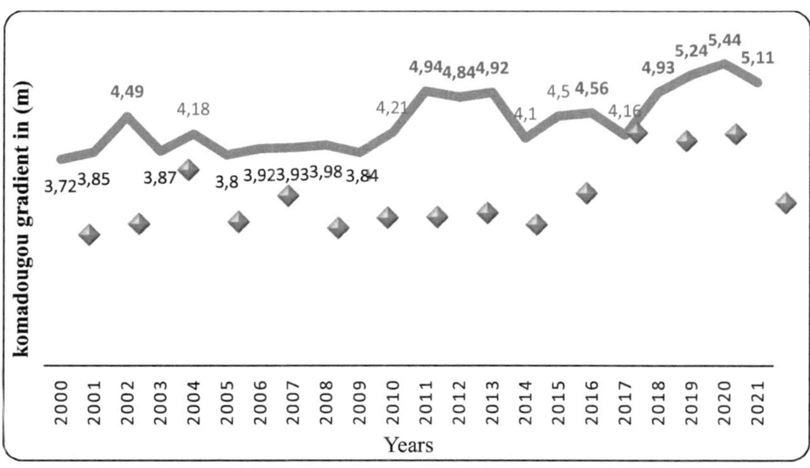

Figure 15 Height reached in m by the Komadougou at the Bagara station (2000-2021)

Source: HRD 2022

Analysis of Figure 15 shows the variation in the level reached by the Komadougou River in m at the Bagara station from 2000 to 2021. The average level reached by the Komadougou over the last 22 years is 4.38m, with a maximum of 5.44m in 2020 and a minimum of 3.8m in 2005. In fact, 2011, 2012, 2013, 2016, 2018, 2019, 2020 and 2021 are the flood years. These years successively exceeded the average level. The latter is one of the causes of recurrent flooding, with adverse consequences for pepper production.

1.12.4 Silting of watercourses

Today, silting plays a key role in the flooding mechanism. The silting up of the Komadougou river takes place during the low-water period, which coincides with the withdrawal of the river's waters. The harmattan wind, which blows from north to south, introduces a large quantity of sand into the river. This reduces the depth of the riverbed and causes it to overflow, leading to flooding. The addition of sand in previous flood years to dam up the water also contributes to the silting up of the river.

1.12.5 Anthropogenic causes

Flooding is a natural disaster, but human activities can exacerbate its intensity and frequency.

1.12.6 Occupation of flood-prone areas by farmers

In the Commune Urbaine of Diffa, the area along the Komadougou Yobé river has been farmed since 1960. Crop fields have proliferated to the detriment of the open forests along the Komadougou river, reducing their original area by 80%. This has caused lateral erosion in the bed of the Komadougou Yobé. The collapse of the riverbanks has contributed to the filling in of the bed and the spreading of water, resulting in an increase in flood-prone areas. Between 1957 and 2007, the area subject to flooding increased by almost 70%. Over the same period, the surface area of ponds increased by 50% (MOUSSA I, 2014). Of the 4 sites in the sample, CBLT is more affected by flooding because it is encircled by the river and therefore located in a high-risk area. This spatial occupation is one of the causes that aggravate flooding in the Diffa Urban Commune.

1.12.7 Impact of flooding on farmland

Analysis of the data shows that the floods have contributed to a reduction in the area under pepper cultivation. Since 2019, flooding on the Komadougou has occurred almost every year. Moreover, the damage caused by each of these floods is often extensive, forcing many growers not to cultivate their areas.

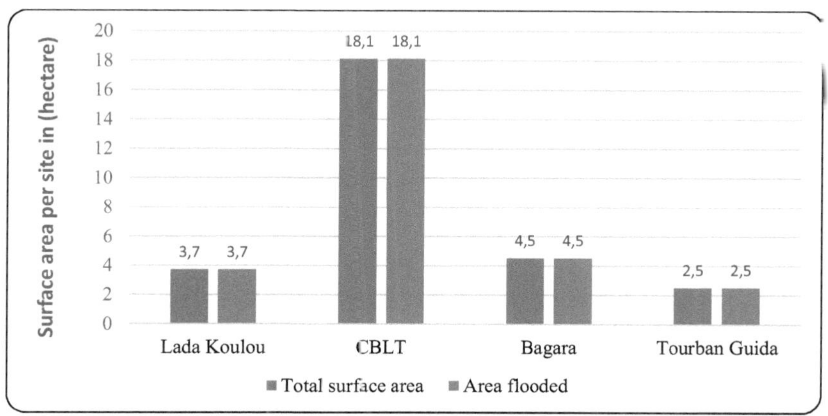

Figure 16 Area affected by flooding

Source: field data, 2022

Figure 16 shows a significant proportion of the area affected by recurrent flooding. Thus 28.8 hectares, or 26.88% of the total area of the sample, were abandoned by producers because of recurrent flooding.

1.12.8 Impact of flooding on production

The impact of flooding on pepper production is very significant in the study area. The survey data made it possible to assess production before the recurrent floods, which gave an idea of the impact of flooding on pepper production.

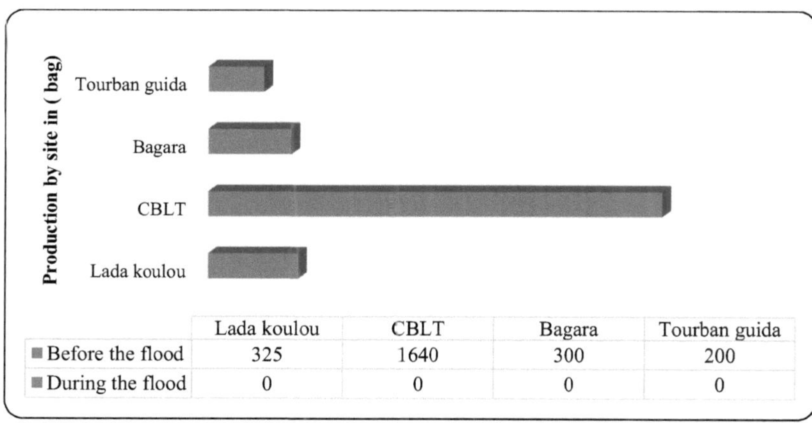

	Lada koulou	CBLT	Bagara	Tourban guida
Before the flood	325	1640	300	200
During the flood	0	0	0	0

Figure 17 Production before and during the sack floods

Source: field data, 2022

Figure 17 shows that the 32% of respondents affected by flooding produced 2,465 bags of peppers before the recurrent floods. Cumulative production at all sites during recurrent flooding was 0 bags. The results show that these floods cause crop losses that can even lead to a complete break in production.

1.13 Impact of restrictive measures on pepper production

The aim here is to tell the story of the transition from one punishment to another for pepper growers, from armed conflict to a state of emergency. At the beginning of February 2015, *Boko-Haram* insurgents launched attacks on the village of Bosso and the town of Diffa. In response to these attacks, the Nigerien government decided on 10 February 2015 to declare a state of emergency throughout the Diffa Region in order to combat the *Boko-Haram* group. And 15 days later, in favour of extending the state of emergency for a period of three 3 months from 24 February 2015.

Three 3 measures have been taken as part of the state of emergency concerning pepper production. These include: a formal ban on the export of peppers to Nigerian markets, the rationing or ban on the sale of hydrocarbons in cans and other containers, and the rationing of the sale of agricultural fertilisers. The analysis therefore concerns only 23% of the sample surveyed, who have continued to develop their farms despite the insecurity and flooding. Out of a total of 107.15 hectares spread across the 4 survey sites, only 28.75 hectares are being farmed despite the security crisis and flooding. The restrictive measures imposed by the state of emergency have had an impact on access to production areas, on the amount of land available for cultivation, on the supply of fuel and fertiliser, and on pepper exports to Nigeria.

1.13.1 Difficult access to production areas

Farmers only have access to 28.75 hectares, or 26.83% of the 107.15 hectares that made up the total area farmed before the crisis. In order to continue their activities, growers have to work in highly militarised areas because of the restrictions imposed under the state of emergency. The farmers interviewed stressed that the security arrangements were an obstacle to access to farmland. To get to their farms, they are subject to checks by the military, especially at the Lada Koulou and CBLT sites.

As farmers have access to their land, it is difficult for them to develop all of their acreage, due to the lack of fertiliser and fuel to power their motor pumps. The state of emergency measures have imposed restrictions on the sale of hydrocarbons and fertiliser. This is the measure most deplored by pepper growers, who are experiencing real difficulties in obtaining fuel for irrigation.

1.13.2 Difficulty in sourcing fertiliser

The measures taken as part of the state of emergency have limited producers' ability to obtain supplies of fertiliser, which is used by terrorists to manufacture explosive devices.

Although it is an important input in pepper production, the sale of fertiliser is now very tightly controlled. As a result, some growers are having to reduce the size of their farms.

1.13.3 Difficulty exporting to Nigeria

Income from the sale of peppers and fish is said to be a source of funding for terrorists. This justifies the ban on exports to Nigeria. This has led to poor sales of peppers, resulting in a significant loss of income for producers. Nigeria accounts for more than 80% of the market for sweet peppers produced in the Diffa region.

1.13.4 State of emergency measures as experienced by pepper growers

It is clear that most of the measures taken under the state of emergency are likely to affect producers' activities. Virtually all the producers surveyed said that they had never been involved or consulted, even when it came to taking measures that directly affected their business (Figure 18). The producers surveyed make no secret of their dismay at the restrictive measures imposed under the state of emergency. They restrict farmers from carrying out their normal activities.

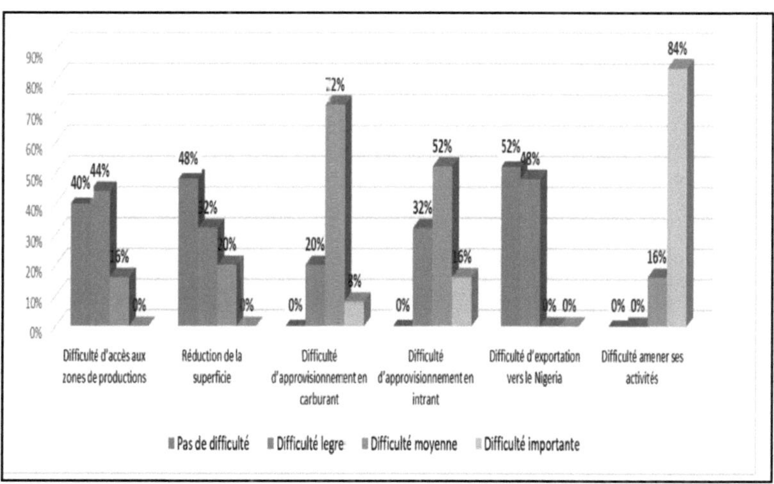

Figure 18 Impact of restrictive measures

Source: field data, 2022

Figure 18 shows that these measures have had a significant impact on pepper production. Thus, 44% of growers have slight difficulty accessing their fields, while 40% have no difficulty accessing their fields and 16% have moderate difficulty accessing their fields.

With regard to the ban on selling hydrocarbons in cans, 72% have average difficulty, 20% of producers have slight difficulty in obtaining supplies of fuel while 8% have major difficulty in obtaining supplies of irrigation fuel.

With regard to the sale of fertiliser, which is highly controlled because of insecurity, 52% of producers have moderate difficulty obtaining supplies of fertiliser, 32% of producers have slight difficulty obtaining supplies of fertiliser and 16% have major difficulty obtaining supplies of fertiliser for production.

These last two measures had led to a reduction in the area of their farms, prompting 20% of producers to make an average reduction in the area of their farms and 32% to make a slight reduction in the area of their fields, while 48% did not reduce their area.

Following the ban on pepper exports to Nigeria, 48% of producers are experiencing slight difficulties in selling their products, while 52% of producers have no difficulty in selling their produce. These analyses show that the state of emergency is having a negative impact on the pepper sector in the Diffa Urban Commune. Indeed, 84% of producers are experiencing major difficulties, while 16% are experiencing moderate difficulties in carrying out their activities during the crisis due to the application of the state of emergency restrictions.

1.14 Local strategies for dealing with the *Boko Haram* security crisis and flooding

The insecurity that accompanied the *Boko Haram* incursions and the escalation of violence that followed the Niger army's counter-offensive, as well as a succession of natural disasters, particularly flooding, have inevitably disrupted the economy of the Diffa region. As a result of the problems associated with the security crisis and the flooding, some pepper growers have adopted new strategies: reducing working hours, changing the cropping calendar, reducing the area under cultivation and mobilising a large workforce.

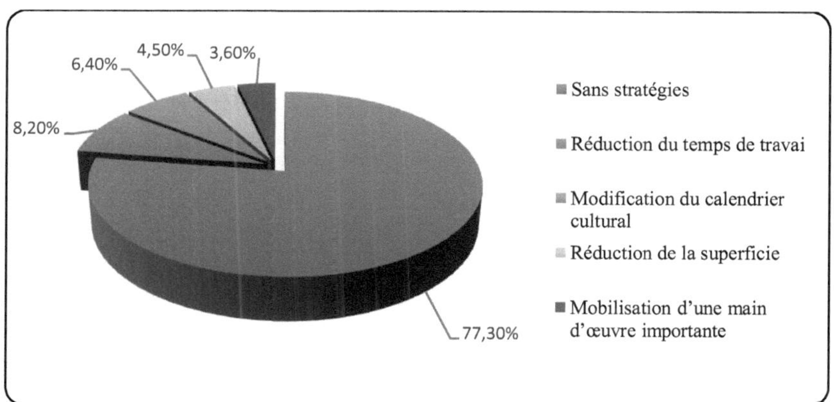

Figure 19 Different adaptation strategies used by producers

Source: field data, 2022

Analysis of Figure 19 shows that 77.30% of growers have no strategy, while 8.20% of growers have a strategy to reduce working time, 6.4% have had to change their cropping calendar, 4.50% have reduced the size of their farm and 3.60% have used a large workforce.

1.14.1 No strategy

The threat of Boko Haram faced by the majority of respondents (77.30%) complicates the implementation of strategies to deal with this security crisis. This is explained by the geographical position of their production areas. They straddle the border between Niger and Nigeria. This area is the scene of intense clashes between the defence forces and terrorists. As a result, growers are unable to put strategies in place to visit their farms, and have had to stop growing peppers altogether.

1.14.2 Changes to the cultivation calendar

This strategy concerns 6.40% of farmers in areas directly affected by the security crisis, but which are still accessible. Farmers have completely changed their cropping calendar. Before the security crisis, land preparation took place in July and August and harvesting began in December. During the security crisis, preparation and sowing took place in May and June, with harvesting in October. This new cultivation calendar is based on the availability of the Komadougou River, which does not allow terrorists to cross.

1.14.3 Reducing working hours in the field

According to the data collected, 8.20% of respondents reduce their working hours in the field. Some pepper-growing areas are under army control, notably the CBLT, Lada Koulou and

Bagara sites. As a result, access to the production areas is restricted to fixed hours, from 9am to 4pm. This is disrupting agricultural work and forcing farmers to reduce their working hours. The measures taken by the government as part of the state of emergency have contributed to the reduction in working hours. Before the *Boko Haram* security crisis, some farmers were working from 5am to 10pm.

1.14.4 Recruiting a large workforce

According to 3.6% of respondents, the use of farm labourers is a strategy that saves time so that work in the fields can be finished as quickly as possible, as *Boko Haram* terrorist attacks can occur at any time. This is why some farmers hire a large workforce for tasks such as preparing plots and transplanting peppers. This strategy is not widely adopted, however, due to a lack of financial resources on the part of growers and a shortage of labour as a result of the security crisis.

1.14.5 Reduction in surface area

As part of this adaptation strategy, 4.50% of producers are reducing their production area. The measures taken by the government as part of the state of emergency, with the rationing, or even banning, of the sale of hydrocarbons in cans, the rationing of the sale of agricultural fertilisers and also the difficulties of access to production areas, have led some producers to reduce the size of their farms. All these factors have contributed to the adoption of this strategy. Farmers now only cultivate small plots of land close to the villages.

1.15   Farmers' coping strategies following the floods

For the past decade, the Urban Commune of Diffa has experienced recurrent flooding, with serious negative consequences for the population and their livelihoods, infrastructure and the environment. Faced with the various risks associated with climate variability - drought, flooding, etc. - which result in a reduction in agricultural production, farmers are developing a number of coping strategies. These include backfilling, dyking, relocating production areas and changing the cropping calendar.

1.15.1 Backfill

Based on the results obtained, backfilling with sand is the technique most used by 90% of growers. Water drainage was the second most popular option (10%).

1.15.2 Submersible dyke

The technique consisted of making barricades out of sandbags of a certain size to line the crops. However, according to the people interviewed, almost all of these dykes gave way in the face of the hazard.

1.15.3 Relocation of production areas

When it comes to limiting the agricultural stakes, farmers tend to leave the lowlands for the uplands, despite the considerable cost of irrigation, as the area requires mini-bores or relaying of motor pumps.

1.15.4 Change of crop calendar

Others are resorting to new strategies such as changing the cropping calendar, crop types and/or using adapted seeds. The cultivation of wheat, barley and tomatoes is now booming, to the detriment of certain crops such as onions and okra.

Partial conclusion

The *Boko Haram* security crisis has had serious repercussions on the economy of the Commune Urbaine de Diffa. These are much more pronounced for pepper production. These repercussions are reflected in lower yields. This has made farmers very vulnerable in this context of insecurity. But in the face of this, coping strategies have been developed to continue developing their plots.

DISCUSSION OF THE RESULTS

This study shows that the *Boko Haram* security crisis has had an impact on pepper production in the urban commune of Diffa. Analysis of the survey results shows that 45% of the producers surveyed stopped growing peppers as a result of the *Boko Haram* security crisis. The data from this study indicate that 49.6 hectares, or 46.29% of the total area, have been abandoned by their owners, and production has fallen from 6,680 to 994 bags of dry peppers, a drop of 85.11%. These results are similar to those of CRA, (2020); FEWS NET, (2018) which show that the security crisis has had an impact on production and also on the situation of producers. The number of producers and the area sown are still 3 to 4 times lower than before the security crisis. According to estimates from the services responsible for monitoring pepper production, of the 8,000 ha of land set aside for pepper production, only 2,000 to 2,500 ha are being developed, with expected production of 2,000 to 2,500 t, compared with an average of 8,000 to 10,000 t. These three studies also confirm our findings Oxfam, (2016); WFP, (2016) and OCHA (2020), the results highlight that nearly 80% of the population earns its livelihood, has for some years also been impacted by insecurity in the border areas with Mali, Burkina Faso and Nigeria. The actions of non-state armed groups will continue to considerably reduce farmers' access to their fields, as well as herders' access to grazing areas.

All the growers interviewed mentioned this as one of the biggest problems caused by the *Boko Haram* security crisis. The state of emergency has resulted in increased militarisation throughout the Diffa region, including the pepper-growing areas. In fact, the state of emergency has resulted in restrictions on access throughout the production zone, with 45% of respondents stressing that lack of access to their production land is a major issue in this security crisis. These results corroborate those of GERAUD M and CHRISTINE R, (2018); ONU, (2018); OCHA, (2018) and MARC-ANTOINE P, (2017) who reach the same conclusions that the state of emergency in the Nigerian states of BYA Borno, Yobe and Adamawa in May 2013, then in the regions of Diffa in Niger in February 2015, Northern Cameroon in July 2015 and the Chadian part of Lake Chad in November 2015 marked a break. Along the Cameroonian, Chadian and Nigerian borders with Nigeria, for example, identity checks were stepped up, the sale of petrol was restricted and motorbike traffic was banned at night, and even during the day in some places; in Diffa, access to fertiliser was rationed because the ammonia contained in urea could be used to make bombs. As a result, the region's farmers ran out of fuel for their motorbike pumps and had to stop cultivating their fields, let alone exporting their produce to Nigeria, their main market.

With regard to the consequences of flooding, the surveys show that 32% of respondents have abandoned their production land since 2019 because of the recurrent flooding that has occurred every year since then, causing considerable damage. Analysis of the survey results shows that 28.8 hectares, or 26.88% of the total area in the sample, has been abandoned by producers because of recurrent flooding. In addition, the damage caused by each of these floods is often extensive, forcing many growers not to develop their land. These are in line with the work of DDA/D, (2019) and ARI M, (2020), the floods have caused significant damage to crops, which are the first and most affected activity on the banks of the river, forming part of the main source of food for riverside households. The local flooding of the Komadougou Yobé in October 2019 invaded and destroyed almost all the irrigated and non-irrigated areas in the zone. Several rice, pepper and onion fields have been flooded. These fields were at the harvesting stage during last year's flood. Any produce not harvested by October has been flooded and destroyed by the waters. And then there are the sunken gardens.

According to the work of the DDA/D, (2019) with a heavy toll of damage loss of life recorded, 358 households affected, 1,075 people affected, 2,877 houses collapsed, 8 cases collapsed,

170.72 ha of market garden crops destroyed, 580.21 ha of rice and pepper crops destroyed, 3 gardens buried, 2 wells damaged, 7 shops collapsed. The loss of cereals is estimated at 300 kg. In addition to the security situation, which has had an impact on production for the past 5 years, the flooding of cultivation plots by the waters of the Komadougou and the pressure of parasites are handicapping the future of this priority value chain.

GENERAL CONCLUSION

This study made it possible to describe the pepper production system and also to analyse the impact of the *Boko Haram* security crisis and flooding on pepper production, and to highlight the various adaptation strategies used by producers to cope with these constraints in the Diffa Urban Commune.

The Komadougou Yobé valley has hydromorphic soil that is ideal for pepper production. There is an abundance of water in the area, with shallow groundwater and surface water from the river lasting more than 7 months of the year. All of which goes to show that this valley abounds in important resources for pepper production.

The vast majority of growers in the study area use their own seed saved from previous seasons, so seed is not a constraint for growers. According to the survey data, 96.36% of the sample use local seed saved by producers. Only 3.64% of producers use seeds bought on the market. For the purposes of this study, the size of farms was used as a criterion for classifying producers. The results of the surveys show that 44% of farmers are small-scale growers, 47% are medium-scale growers and 9% are large-scale growers. Medium and large growers therefore represent more than half of the sample, which means that yields are high at this level.

The two varieties used by producers in the Commune Urbaine de Diffa, *Kangadi N'Glaro* (sheep's horn) and *Mouri Koro* (donkey's dung) seem to be well adapted to the production zone. The use of inputs and pesticides is crucial to pepper production, as it determines yield. All those surveyed said they used pesticides, and 68.18% of the sample used chemical fertilisers, while 22.72% used both chemical and organic fertilisers.

Peppers are easy to dry and store. Properly dried, they can be stored for several years without significant loss. What's more, storing peppers poses no problems. When dry, they are not subject to insect attack. These results therefore confirm the first hypothesis of this study, which states that the pepper growing system normally allows good production.

With regard to the impact of the *Boko Haram* security crisis and the recurrent floods. The *Boko Haram* security crisis led to a reduction in the area under production and a drop in pepper production. Out of a total of 107.15 hectares divided between the 110 growers surveyed, 49.6 hectares or 46.29% of the total area has been abandoned because of the security crisis. On the other hand, 28.8 hectares (26.88%) have been abandoned by growers because of recurrent flooding. The 75 growers affected by the crisis had 6680 bags of dry peppers before the crisis and 994 bags of dry peppers during the security crisis (a drop of 85.11%). Producers affected by the floods produced 2,465 bags of peppers before the recurrent floods. While production during the recurrent floods was 0 bags. So the *Boko Haram* conflict, the related security measures and the floods have led to a significant reduction in pepper production. This cor firms the second hypothesis of the study, which states that the security crisis and the floods le 1 to a drop in pepper production.

Despite the security crisis and recurrent flooding, growers have adopted various adaptation measures to deal with the conflict and recurrent flooding. In addition, to deal with the various threats, farmers say they have developed a number of strategies. For example, 8.20% of farmers have reduced their working hours, 6.4% have had to change their cropping calendar, 4.50% have reduced the size of their farm and 3.60% have mobilised a large workforce. This confirms the third hypothesis of the study, according to which farmers have adopted strategies to cope with security constraints and flooding.

In terms of prospects for the doctoral studies, the analysis will focus on the impact of insecurity on land tenure in the rural communes of Chétimari, Bosso and Gueskerou/Diffa Region.

BIBLIOGRAPHY

1)   **ABDOU A et *al*** (2005). *Etude sur l'approfondissement du diagnostic et l'analyse des systèmes de production agro-sylvo-pastoraux dans le cadre de la mise en œuvre de la stratégie de développement rural.* Diffa Region, pp 51-59.

2)   **ABDOULKADRI L.** (2014). *Contribution à l'étude de la dynamique de l'élevage pastoral au Niger : cas de la région de Diffa.* PhD thesis, University of Liège - Gembloux, 8p.

3)   **ABDOURAHAMANI M.** (2011). *Cropping systems in the Boultoungour polder and impacts on population food security (Lake Chad, Eastern Niger).* Diplôme d'études Approfondies dissertation, UAM, 48p.

4)   **ABHAS K, ROBIN B. and JESSICA L.** (2012). *Cities and floods: a guide to integrated urban flood risk management for the 21st century*, 63 p.

5)   **ADAMOU M.** (2019). *Cross-border insecurity in West Africa: the case of the Niger-Nigeria border.* Political Science, Université Côte d'Azur, pp 87-89.

6)   **ALEXANDRE M, NEELAM V and STEPHEN M.** (2015). "*Meeting the challenges of stability and security in West Africa: Executive summary.*". Developing Africa Collection, 14p.

7)   **ANTOINE K.** (2015). *Etude de la contamination du poivron (Capsicum L., 1753) par les produits phytosanitaires,* Mémoire pour l'obtention du Diplôme de Master, Chimie de l'Environnement UNA pp 2-16.

8)   **ARI M.** (2020). *Socio-economic and environmental impacts of flooding of the Komadougou Yobé in the Urban Commune of Diffa: case of the 2019 local flood.* Master's thesis, University of Dosso, 58p.

9)   **BENOIT S, and ADAPTE S, AGRHYMET** (no date). Le Sahel face aux changements climatiques enjeux pour un développement durable, pp 9-10.

10)   **BELKHIRI A and SEFIH F.** (2016). *Bio-ecological study of the parasitic complex of greenhouse pepper aphids in the Mostaganem region.* Master's thesis in agronomy, 2p.

11)   **BIGA I.** (2008). *Etude de la biodiversité végétale des milieux de culture du poivron (Capsicum annuum L.) : cas des terroirs villageois de Gueskérou, Diffa et Tam (Région de Diffa, Niger).* Master's thesis, University of Niamey, 16p.

12)   **BOUDOUKHA A, BOUMESSNEGH A.** (2012). Les inondations en Algérie genèse et impacts (cas de la ville de Biskra) Sud Est Algérien, 442 p.

13)   **CANDY J.** (2006). *Effet de la durée de compétition des mauvaises herbes sur la culture du poivron (Capsicum annuum), Mémoire du diplôme d'ingénieur Agronome, option : Production Agricole et Transformation des denrées.* Université Notre Dame d'Haïti, 57p.

14)   **DIFFA REGIONAL CHAMBER OF AGRICULTURE.** (2010). *Note d'information / Filière pêche n°1,* 5p

15)   **DIFFA REGIONAL CHAMBER OF AGRICULTURE.** (2016). *Le poivron rouge de Diffa éléments techniques et économiques pour la culture,* 1p.

16)   **REGIONAL CHAMBER OF AGRICULTURE.** (2020). *Monitoring the red pepper market in Diffa: evaluation and prospects for the Manga red gold,* 7p

17)   **CNEDD.** (2019). Mapping the vulnerability of agropastoral activities in the regions of Niger as part of the PDIPC Project 70p.

18)   **RURAL DEVELOPMENT COMMISSION.** (2006). *Contribution to the review of the poverty reduction strategy,* 62p.

19)   **COHAND J.** (2007). La petite irrigation privée dans le sud du Niger : potentiels et contraintes d'une dynamique locale le cas du sud du département de Gaya. Dissertation, University of Lausanne, 7p.

20)   **CRDEL.** (2010). *Flooding in Greater Cotonou: human factors, population vulnerability and control and management strategies,* final report, 82p.

21)   **DOSSOU J.** (2015). *Health impacts of floods on populations in Cotonou, Benin: Case of the 9th arrondissement of Cotonou.* Mémoire pour l'obtention de Licence professionnelle en GEn/EPAC/UAC, 37p.

22)   **EUROPEAN ASYLUM SUPPORT OFFICE EASO.** (2018). *Country of origin information report Nigeria - individuals targeted,* 23p.

23)   **FAO.** (2016). *An assessment conducted by FAO in October 2016 in Borno, Yobe and Adamawa States, Nigeria,* 13p.

24)   **FAO.** (2017). *Mitigating the impact of the crisis and strengthening the resilience and food security of conflict-affected communities,* Rome, 6p.

25)   **FEWS NET.** (2018). *Sécurité alimentaire situation de Stress (Phase 2 de l'IPC) en zone pastorale,* 7p

26)   **GERAUD M AND CHRISTINE R.** (2018). The Lake Chad region facing the Boko Haram crisis: interdependencies and vulnerabilities of a Sahelian hinge, 20p.

27)   **GERAUD M and MARC-ANTOINE P.** (2018). The Lake Chad region in the face of Boko Haram, 13p.

28) **HADIZA K.** (2014). *Impacts of variations in the level of Lake Chad on the socio-economic activities of fishermen in the Niger part.* PhD thesis, Abdou Moumouni University of Niamey, 55p.

29) **HADIZA K.** (2019). *Study on fishing and the fish sector in the Diffa Region.* Ministère de l'agriculture et de l'élevage, Programme de développement de l'agriculture familiale (ProDAF), Unité Régionale de Gestion du Programme (URGP) de Diffa, 123p.

30) **HAMADOU I and DOMINIQUE B.** (2013*). Les inondations à Niamey, enjeux autour d'un phénomène complexe,* p. 295-310. Available online at https://doi.org/10.4000/com.6900 (consulted on 05/03/2022).

31) **HAMANI O and ABDOURAHAMANE O.** (2017). La gestion humanitaire des inondations dans une commune de Niamey, Report 10 p

32) **Http://Www.**Wikipedia.com: varieties of sweet peppers of the species Capsicum annuum with very large fruits, consulted on 06/02/2021 at 20:42).

33) **INSTITUT NATIONAL DE LA STATISTIQUE INS.** (2016). *Diffa regional statistics yearbook. Period from: 2011 to 2015,* 24p.

34) **INSTITUT NATIONAL DE LA STATISTIQUE INS.** (2018). *Statistical yearbook 2013 - 2017,* 208p.

35) **INTERNATIONAL CRISIS GROUP.** (2017). *Boko Haram in Chad: beyond the security response Africa Report N°246,* 23p.

36) **ISSA Y.** (2005). *Les transports terrestres dans le processus de désenclavement et d'intégration du Niger dans la sous-région ouest-africaine : L'exemple de la route nationale n°6,* Master's thesis, UAM 73p.

37) **LE COZ M.** (2010). *Hydrogeological modelling of heterogeneous deposits: the Komadougou Yobé alluvium (Lake Chad basin, south-eastern Niger).* PhD thesis, Université Montpellier II, 14p.

38) **Lucile W.** (2010). Flooding in West African cities: Diagnosis and elements for strengthening adaptation capacities in Greater Cotonou, 90 p.

39) **MALIKI** et *al (*2006). Market study of agro-pastoral products in the Diffa region. Final report, 123p.

40) **MARC-ANTOINE P.** (2017). "Nigeria, Boko haram and the migration crisis". *Outre-Terre* (No. 53), 180p.

41) **MARIAMA O.** (2013)**.** *Plan national de contingence multi risque Niger (PNC_MP).* Rapport d'activité, 17p.

42) **MINISTRY OF AGRICULTURE AND LIVESTOCK**. (Undated). *Projet promotion des exportations agro-pastorales, étude sur la facilitation du commerce couvrant les filières agricoles*, ppeap, pp 35-38.

43) **MOUSSA I.** (2014). *Dynamiques érosives et des états de surfaces dans la partie nigérienne du bassin du lac Tchad.* PhD thesis, Abdou Moumouni University, Niamey, 15p.

44) **UNITED NATIONS.** (2018). *Report of the Special Rapporteur on the human rights of internally displaced persons on her mission to Niger*, pp 8-10.

45) **OCHA.** (2017). *Lake Chad Basin: Overview of the crisis*, 2p.

46) **OCHA.** (2020). *Overview of humanitarian needs Niger*, 11p.

47) **OUMAROU H.** (2005). "*Comportement du poivron sur sol salin aux bords de la Komadougou : Cas du site de Chétimari-Gréma-Artori*". Mémoire ITA4, FA/UAM, 55p.

48) **OUSMAN Y.** (2013). *Etude économique de la culture du poivron dans les exploitations agricoles familiales de la région de Diffa, cas du village de Kayowa.* Mémoire de fin de cycle Agronomiques, Option Economie Rurale UAM, pp 18-21.

49) **OXFAM**, (2016): *Income market systems for smoked fish and dried red pepper Diffa Region, Eastern Niger*, 8p.

50) **PDES.** (2012-2015). *Economic and Social Development Plan*, 28p.

51) **PIERRE-FRANÇOIS P and SALIFOU K.** (2005). "*Etude de l'impact de 'a production et de la commercialisation du poivron dans la région de Diffa au Niger*", final report, 59p.

52) **REPUBLIC OF NIGER.** (2003). Office of the Prime Minister, *rural development strategy*, pp 5-6.

53) **REPUBLIC OF NIGER.** (2014). Department of diffa , *communal development plan (PDC)*, pp 24-28.

54) **REPUBLIC OF NIGER.** (2015*). Plan de développement régional de diffa 2016-2020*, 24p.

55) **REPUBLIC OF NIGER.** (2015). Ministry of Agriculture, *Stratégie de la petite irrigation au Niger (SPIN)*, 24p.

56) **REPUBLIC OF NIGER**. (2006). Ministère du développement agricole, *Processus d'élaboration et de mise en œuvre du Code Rural au Niger*, 5p.

57) **REPUBLIC OF NIGER.** (2020). Ministère de l'équipement, *Etude d'Impact Environnemental et Social des travaux d'aménagement, de bitumage et de réhabilitation des voiries de la ville de Diffa dans le cadre du Programme Diffa N'Glaa d'environ 25 km* rapport définitif, 33p.

58) **REPUBLIC OF NIGER.** (2018). Diffa region, *Commission régionale d'aménagement du territoire,* CRAT pp 98-117.

59) **REPUBLIC OF NIGER.** (2012). Région de Diffa, *Plan de développement économique et social, Bilan Diagnostic* PDES Diffa, 107p.

60) **REPUBLIC OF NIGER.** (2000). *Plan national de l'environnement pour un développement durable* (PNEDD), 10p.

61) **REPUBLIC OF NIGER** (2016). WFP, *food security assessment in complex emergencies in the Diffa region,* pp 4-57.

62) **REPUBLIC OF NIGER.** (2017). Ministère du plan, *Stratégie de Développement Durable et de Croissance Inclusive* 12p.

63) **RESEAU NATIONAL DES CHAMBRES D'AGRICULTURE DU NIGER.** (2010). *"Etude de l'impact de la production et de la commercialisation du poivron dans la région de Diffa au Niger",* 3p.

64) **SAGAGI M and THORBUM A.** (2019). *Discussion paper on cross-border trade and regional economy.* Deuxième réunion de Niamey, 16 au 18 juillet, 2p

65) **SIDDHARTH K al** (Undated). The economic consequences of conflict, 25p.

66) **TRISTAN N.** (2004). *Contribution à la stratégie de sélection de génotypes de piments (capsicum annum, L) adaptés aux conditions tropicales chaudes et humides.* Agricultural engineer at the Ecole Nationale d'Agriculture de Thiès, Senegal, 63p.

67) **TARCHIANI V, et** *al* (2021). Floods in Niger 1998-2020. ANADIA2 Project Report, 15p.

68) **UNHCR,** (2017). Site de Boudouri, Commune de Chetimari, Diffa region, Niger, weekly report, 2p

69) **VICTORIA O.** (2015). http://Www.Femininbio.com/alimentation/conseils-et-astuces/vertus-bienfaits-du pepper 54209, consulted on 4/02/2021 at 18h27.

70) **ZITOUNI D and DOUAR K.** (2017). *Bioecological study of the auxiliary fauna of pepper aphids under glass.* Master's thesis in agronomy, 2p.

Appendix 1: Individual questionnaire sent to producers

A-IDENTIFICATION
1. First name and surname of respondent :
2. Age : sex : Male Female
3. Place of birth :
4. Ethnic group :
5. Survey date: .... Day...... me.......
6. Survey website :
7. Level of education

Primary
Secondary
Superior
Koranic
No
8. Marital status

Single
Married
Divorced
Widowed

B - LAND TENURE
9. Is there a land manager at your site?
If so, who is he? ....
If not, why not? ....
10. How are plots of land acquired?

Heritage		Rental		
Purchase		Pledge		
Don		Loan		

11. Who authorises the cultivation of new land?
-The land chief
-The village chief
-The district manager
-others.
12. Are there any associations, NGOs or projects involved in land management?
If yes, which ones..
If not, why.................................... ....

C- THE SECURITY CRISIS / FLOODING AND PEPPER PRODUCTION
13. Have you been personally affected by the BH/ security crisis or the flood?
    Yes No
    If not, maintenance stops.
14. What activities are affected by the BH security crisis/flooding...........?

15. Did the security crisis/flood cause you any damage? If so, what was the nature of the

    damage ....?

16. Do you think the BK security crisis/flooding is the main cause of this damage? If so,

    how? ....

17. Can you give us an estimate of the cost of this damage?   If so, please

In Naire................................. In FCFA..

18. Since the start of the BH safety crisis, has the total surface area of your farm

    increased?

     Augmented

    Decreased

    Stagné

    Why? ....

19. Did you abandon your field during the insecurity/flood?
    Yes
    No
    If yes, how long and why.....................................
20. Did you have access to your field during the BH security crisis? yes no
    If yes how if no why ....
21. What measures do you propose to minimise the scale of this BH security
    crisis?...
    D-RETURN
22. How many hectares do you grow each year? No. of hectares
23. On average, how many bags of peppers do you harvest each year?
24. How many bags of peppers do you harvest before the crisis/flood?
25. How many bags of peppers do you harvest during the crisis/flood?
26. Assessment of pepper yields during the crisis/flood
    Increase
    Constance
    Decrease
E-PEPPERS AND PRODUCTION METHODS
27. What type of soil do you grow peppers on and why....................?
28. Do you fallow? yes no
 If so, for how long and how important is it for your crops........................
29. What time of year do you start growing your crops................................
30. What time of year do you start harvesting ....
31. What are the different procedures for developing land for pepper production
    ....?
32. Do you make amendments?
 Yes No
 If so, which one do you use?
-organic fertiliser
-chemical fertiliser
other
 If not, why not? ....
33. Are the seeds local?
Yes No
If not, where do they come from ....?
34. How do you go about acquiring seeds ....?
35. Do you use production inputs?
Yes
No
If so, which ones and why ....

If not, why...
36. How and from whom do you source your inputs? ....
37. How are your crops supplied with water?
-Rain
 irrigation
 Watering
 other
38. Why do you grow peppers?
- Consumption
National Sale
International
39. Do you encounter any difficulties during production?
Yes No
If so, which ones and when? And how do you solve them ....

F-PEPPERS AND STORAGE

40. Do you keep the pepper?
Yes No
If so, which part do you keep and why? What method do you use? ....
If not, why don't you keep it? ....
41. Are there any conservation problems?
If yes, which ones

....
42. What solutions have been found..
43. Do you grow several varieties?
    If so, what are they?
44. How do you distinguish between these varieties?
    The shape
    colour

Impact of restrictive measures on pepper production in the CUD (state of emergency measures)		
Ban on going to the field (state of emergency measures)	No ban Slight prohibition Medium ban Important prohibition	
Loss of land from pepper production (state of emergency measures)	No losses Slight loss Average loss Significant loss Total loss	
Difficulties in supplying hydrocarbons in cans and other containers for watering (state of emergency measures)	No Difficulty Light difficulty Medium difficulty Major difficulty	
Supply difficulties agricultural inputs (state of emergency measures)	No difficulty Light difficulty Medium difficulty Major difficulty	

Formal ban on exports of peppers to Nigerian markets (state of emergency measures)	No ban Slight prohibition Medium ban Important prohibition	
Ban on motorbike traffic affects farmers (as motorbikes allow farmers to get to the fields easily)	No difficulty Light difficulty Medium difficulty Major difficulty	
Difficulties in carrying out its activities during the security crisis	No difficulty Light difficulty Medium difficulty Major difficulty	

<u>STRATEGIES AGAINST INSECURITY AND FLOODING</u>

45. Can you describe the local strategies you have adopted to continue production despite the insecurity? ....
46. Given the various strategies you are currently developing to deal with insecurity, what strategies do you plan to implement in the future? ....
47. Have you received any strategies from the local authorities to minimise the scale of this crisis? Yes No
    If yes, which ones..
48. And how do you rate the effectiveness of these strategies (the state of emergency measures)?
    Very sufficient
    Quite sufficient
    Insufficient
49. Can you tell us how many of your surfaces have been affected?

Short-term strategies	Long-term strategies

Appendix 2: Interview guide for technical services, customary authorities and other institutions

<u>**IDENTIFICATION**</u>
1.   Last name and first name of respondent: ....
2.   Position of responsibility: ....
3.   Date of interview: ....
4.   Department concerned: ....
5.   File No.: ....
<u>**LAND TENURE FOR PEPPER CULTIVATION**</u>
6.   How are plots of land for pepper production acquired in Diffa CU?
....
Can everyone access the land? yes no
If not, why this distinction..

7.   Who grants exploitation rights?

Local authorities (mayor)	
Customary authorities (canton or village chief)	
Others to be specified	

What are the most recurrent land-related conflicts?...
What are the sources of these conflicts?
....
Who settles these disputes?
....

   <u>**BOKO HARAM INSECURITY**</u>
8.   In your opinion, what is at the root of BH insecurity in the CUD?
....
9.   What do you think of the lack of safety at BH?
....
10.   What do you see as the advantages and disadvantages of this security crisis?
....
11.   How do you rate the effectiveness of the measures taken by the authorities to minimise the scale of this crisis?
   -Very sufficient
   -Quite sufficient
   -Insufficient
12.   What proposals do you have to combat BH insecurity in your CUD?.........

THE IMPACT OF THE CRISIS ON THE PEPPER INDUSTRY
13.   What impact has the security crisis had on the pepper industry?
....
14.   How has pepper production evolved during the security crisis?
Increase
Constance
Decrease
15.   What are the constraints involved in producing peppers?

Insecurity constraints	Natural constraints (flooding)

<table>
<tr><td>

</td><td>

</td></tr>
</table>

STRATEGIES TO COMBAT INSECURITY AND FLOODING

16.   What strategies have farmers adopted to cope with insecurity? ....

17.   What strategies have farmers adopted to cope with flooding? ....

18.   How do you see pepper production in the coming years?
How can it be improved?   Strengthen it? ....

More
Books!

info@omniscriptum.com
www.omniscriptum.com
OMNIScriptum

Printed by Books on Demand GmbH, Norderstedt / Germany

Antonio Moreira

Acupuncture and Probiotics for Constipation

Antonio Moreira

Acupuncture and Probiotics for Constipation

Evidence of the effects of Acupuncture and Probiotics on Constipation

ScienciaScripts

CONTENTS

CHAPTER 1

ACUPUNCTURE AND PROBIOTICS

Two fascinating and enigmatic phenomena. Acupuncture in its main branch uses a piece of metal (needles) to have physical and biochemical effects. Probiotics, bacteria that produce chemical substances to nourish and defend the brain-intestine axis. Metal and bacteria? Yes, exactly! How can this happen? For acupuncture, the body is equipped with a system like a river that runs through mountains and valleys until it reaches the sea. On its way, it divides into channels. Water can represent the energy that follows its path day and night. What it contains is blood with oxygen and various types of life. When the flow of water (energy) is interrupted, it creates stagnation or deficiencies. If maintenance is neglected, toxicities will result. The needles act like a backhoe unblocking the channel. At the same time, like a telegraph, it sends a message to the farmer asking for supplies to nourish the community. The farmer, figuratively speaking, is the brain.

For Probiotics, they play the role of soldiers guarding a barracks. The greater the number of soldiers and their quality, the better the place is protected. Take, for example, the vaginal flora, which protects the organ through competition. If the command is affected, psychobiotics come into play to act on the emotional state. Like a worker bee producing royal jelly for the queen, probiotics produce nutrition for the neurotransmitters. According to the 5 Elements of Chinese medicine, it is Earth that dominates Water. A lot of earth absorbs water. There can be counter-dominance. Too much Water floods the Earth. In the body, probiotics protect the intestines. If there are no probiotics, pathogenic bacteria take over and dysbiosis occurs. Isn't that fascinating and enigmatic?

Straw to stimulate the imagination

For lovers of planet earth and those seeking longevity, I'm going to give you a flavour of the wonderful news you'll find in the body of the text. Recent studies have shown that probiotics reduce the excessive generation of reactive oxygen species. With ageing, the body becomes less efficient at managing environmental stress, leading to a

myriad of physiological imbalances and mitochondrial damage. Probiotics can promote microbiota maintenance and contribute to longevity (Westfall et al. 2018).
Mitochondrial refers to the energy plant (mitochondria)
Reactive oxygen species are the famous free radicals that cause premature ageing.

Dear readers, before the advent of the internet, a book came into our hands out of date, and the author wrote what he thought. Now, in modern times, a document like this is based on evidence from controlled studies and reviews of controlled studies. The references for this document were taken from the science website PubMed, which is also a reference base for master's and doctoral programmes. In this way we have a text that is rich in information and reliable.

I've been meaning to write about this subject for some time. A subject that deals with a syndrome that doesn't choose an age or time to manifest its symptoms of physical and psychological pain. I have written with diligence and responsibility so that the information described is reliable.

INTRODUCTION

Of all the things that are considered perfect, the functioning of the human organism is sublime. Our body is a fascinating piece of engineering, created and assembled by hand. All the organs, tissues and mind are connected. In genetic inheritance, the body with a compromised prenatal essence, and the postnatal essence acquired through inappropriate behaviour after birth, disharmonises the composition. The human organism is equipped with an energy network that connects the exterior and interior. From the outside comes the nutritional essence of the air and combines with the nutritional essence of food to form the energy that drives life.

In the digestive and absorption system, it is given by peristaltic movements and enzymatic processes in Western scientific studies. Traditional Chinese medicine, on the other hand, has a simpler physiological explanation. Everything depends on the movement and quality of IQ (energy).

The intestines, due to their connection with the brain and the production of neurotransmitters, can be considered a second brain. When the intestine-brain interdependence is disturbed, the entire physiological system is affected.

As an intestinal syndrome, constipation is related to physical, biochemical and psychological factors. Acupuncture, enigmatic and fascinating, is a therapy that can effectively treat constipation. Probiotics, like acupuncture which has a long history, play an essential role in maintaining human health. Acupuncture and probiotics form a remarkable duo in the treatment of constipation and all intestinal syndromes.

CONSTITUTION

Constipation is a syndrome that affects everyone from newborn babies to the elderly. It causes great suffering for the sufferer and their carers. It affects quality of life and can last a lifetime. Complaints are common in doctors' surgeries, gyms and all areas of health care. Changes in health range from abdominal pain to psychosocial problems. In the newborn, it causes pain, crying and distress for the mother. In children, the pain symptoms cause fear of defecation. In adolescents, it can lead to mood swings and school problems. In adults, it can lead to absenteeism from work due to the discomfort it causes, such as body pain and depression. Women suffer more because of their social behaviour - they don't defecate just anywhere. They are affected in terms of body composition, skin health and emotional state.

In Chinese medicine, the organ responsible for the free flow of emotions is the Liver. As a result, women experience frequent changes in their emotional states due to hormonal cycles, and suffer from stagnation and deficiencies related to the Liver, leading to various disharmonies, including constipation. In the elderly, due to the lack of motor activity and the constipating effects of certain drugs, they also suffer from the discomfort of constipation. Constipation has a variety of causes and so far is not well understood. Nowadays, due to the number of complications that constipation causes, treatment is becoming more important. However, many professionals who don't appreciate the symptoms of constipation still prescribe laxatives as the main course of action, which doesn't solve the problem and causes even more discomfort.

INTEREST IN CONSTIPATION STUDIES

The interest is due to the dissatisfaction of a large number of women with conventional treatments. I've been observing this for over two decades. I'm referring to women in particular because they are most affected. There are several situations that favour constipation in women. In a developing country that doesn't cater for women's needs, they are forced to adapt to existing conditions. As there are more of them in all sectors of society, they avoid drinking water so as not to have to go through the hassle of not having enough toilets to cater for them. For the majority of women, the workday is double, they work at their jobs and when they get home they have to deal with household chores, children, etc. They lack time for physical activity and the act of defecation itself. The human organism adapts or creates addictions according to each person's behaviour. When the urge to defecate arises and this urge is stopped, the body creates means of adaptation. It absorbs most of the intake, contributing to an increase in the percentage of fat, or because of the discomfort generated, it is forced to resort to laxatives, causing cramps and diarrhoea. In consultations, constipated women report that defecation is torture because of the pain and anal bleeding. Digital aids - the use of one's own finger to remove faeces impacted in the rectum - have a high prevalence. Due to the existing suffering, body pain, abdominal distension, depression and reduced sexual libido, I am writing so that we can have better clarity on the treatment of constipation.

WHAT TO DO TO PREVENT IT?

We need to be more careful about prevention. When we work on prevention, even if the disease appears, it comes more gently and is easier to treat. Lack of physical exercise, failure to manage the mind, low water intake, too few hours of sleep and an inadequate diet all contribute to the disease state.

Prevention is necessary from the prenatal essence, which is acquired genetics, through to the postnatal essence, which is understood as post-birth behaviour, until the end of the body's existence.

It's still not well understood, but today we know from studies of epigenetics, nutrigenomics and nutrigenetics that diet has a direct or indirect relationship with genetics. As such, diet can affect life from conception to late adulthood.

Women who want to conceive should start taking care of their health well before pregnancy. Pleasurable exercise, a diet with food that comes directly from the waters and the earth, and restful sleep. During pregnancy, care needs to be redoubled and more disciplined. Diet with supplementation including probiotics, moderate exercise and acupuncture for nausea and anxiety.

During childhood and adolescence, parents should encourage their children to take part in physical activity and eat a diet that nourishes and does not damage the body. Avoid contaminating and inflammatory foods as much as possible. Enrich their minds with affectionate behaviour.

For adults, use the arts of doubt and criticism in the face of adverts, using intelligence and wisdom at all times to identify "professionals" who use science to sell medicines. Have nutrition and exercise as the main remedies. Nutrition that is based on anti-inflammatory, antioxidant and antacid dietetics. As well as on the physiology and philosophy of oriental dietetics, which follows the principles of the 5 Elements and the energetic relationship of Yin/Yang. Exercises to stimulate neurotransmitters and general well-being. As a complement or treatment, use acupuncture and probiotics.

For young children and bedridden elderly people, acupuncture in its various tools such as moxibustion, manual stimulation and many others, probiotics and trace elements within the principles of quantum physics.

And finally, beware of popular prescriptions that are not based on evidence or long-standing observations.

DEFINITION OF CONSTIPATION

Constipation is a disorder of the gastrointestinal tract, which can result in infrequent stools, difficulty passing stools with pain and stiffness. The pathogenesis is multifactorial, focusing on genetic predisposition, socioeconomic status, low fibre intake, lack of adequate fluid intake, lack of mobility, disturbances in hormonal balance, side effects of medication or body anatomy, etc.(Forootan et al. 2018)

Chronic constipation is a persistent and common condition that affects many patients worldwide, presenting a significant economic burden and resulting in substantial utilisation of health services. In addition to infrequent bowel movements, the definition of constipation includes excessive straining, the sensation of incomplete evacuation, unsuccessful or delayed attempts to defecate, the use of digital manoeuvres to evacuate faeces, abdominal bloating and rigid stool consistency. After excluding secondary causes of constipation, chronic idiopathic or primary constipation can be classified as functional defecation disorder, slow transit constipation, and irritable bowel syndrome with predominant constipation. These classifications are not mutually exclusive and there is significant overlap (Sharma and Rao, 2017).

Constipation is characterised by infrequent or difficult to pass bowel movements accompanied by straining during defecation, or a feeling of incomplete evacuation. It is also a subtype of irritable bowel syndrome in which constipation and altered bowel habit predominate, often together with recurrent abdominal pain and abdominal distension (Mearin et al. 2017).

Constipation is a problem that affects all ages. Chronic constipation usually has an insidious onset over many years, often dating back to childhood (Mccrea et al. 2008).

EASTERN MEDICINE DEFINITION

Functional constipation can be attributed to defecation difficulties in traditional Chinese medicine (TCM), which are considered to be caused by inadequate diet, emotional disturbances and invasion by exogenous pathogens. According to the different pathogenic factors, defecation difficulty can be classified into the categories of Qi deficiency constipation, Blood deficiency constipation, Yin deficiency constipation, etc. These disorders are attributed to dysfunction of the intestinal transit function and insufficient intestinal fluid (Xu et al. 2014).

Pathologies in oriental medicine are referred to as energetic imbalances in a particular element, or organs and viscera (Zang-Fu). In the case of constipation, it is a deficiency in the Earth element whose organ is the Spleen. The Spleen is considered to be the officer of transformation and transport. Among many functions, the movement of material in the process of digestion through the Small and Large Intestines and, finally, the movement of material to be excreted through the Large Intestine and out of the anus are largely under the control of the Spleen (Hicks et al. 2007).

CAUSES OF CONSTIPATION

The pathogenesis is multifactorial, focusing on the type of diet, genetic predisposition, colonic motility and absorption, as well as behavioural, biological and pharmaceutical factors. In addition, low fibre intake, inadequate water intake, sedentary lifestyle, irritable bowel syndrome, failure to respond to the urge to defecate and slow transit have been associated with predisposition (Forootan et al. 2018).

Constipation comes in different types, including constipation with no apparent cause, which should be investigated carefully. According to Black and Ford 2018, chronic idiopathic constipation is one of the most common gastrointestinal disorders, with a global prevalence of 14 per cent. It is more common in women and its prevalence increases with age. There are three subtypes of chronic idiopathic constipation: dyssynergic defecation, slow transit constipation and normal transit constipation, which is the most common subtype.

In childhood, the underlying pathophysiology of functional constipation is multifactorial and poorly understood. Factors that may contribute to functional constipation include pain, fever, dehydration, food and fluid intake, psychological problems, toilet training, medications and family history of constipation (Diaz and Mendez, 2018).

Chronic idiopathic constipation is characterised by infrequent defecation, difficulty or pain during defecation without an identifiable organic cause, such as physiological, anatomical, radiological or histological. Although this is a common problem in children, an underlying cause is identified in less than 5% of constipation cases (Diaz and Mendez, 2018).

Constipation is a heterogeneous, polysymptomatic and multifactorial disease. Acute or transient constipation can be due to changes in diet, travelling or stress, and secondary constipation can result from drug treatment, neurological or metabolic conditions or, rarely, colon cancer (Rao et al. 2016).

Constipation, a common symptom among stroke patients that leads to increased morbidity and mortality, has been reported to be associated with an increased length of hospital stay, worse neurological prognosis and death. The incidence of constipation in patients with steatosis has been reported to range from 29 per cent to 79 per cent. In addition, patients with post-stroke constipation are more likely to experience medical complications such as pneumonia, urinary tract infection, upper gastrointestinal bleeding and recurrent stroke (Wen et al. 2018).

According to Taba Taba Vakili et al. 2015, a high-fat diet activates the ileal brake, which delays gastric emptying and intestinal transit, decreasing overall gastrointestinal transit and predisposing to constipation.

PATHOPHYSIOLOGY OF CONSTIPATION

Functioning normally, the colon absorbs fluids and transports waste to the rectum through the repetitive and periodic contractions of peristalsis, mediated mainly by serotonin or 5-hydroxytryptamine (5-HT). Sodium is actively reabsorbed through active transport channels; water by osmosis. Colonic secretion is mediated by chloride channels and results in a net reabsorption of electrolytes and fluids. The rectum eventually distends, resulting in the urge to defecate and associated contractions via the rectal sphincter. Constipation represents a disruption of these normal mechanisms. The causes can be primary (colonic or anorectal dysfunction) or secondary (related to disease or medication). Factors contributing to constipation can include disruption of normal motility, excessive dryness of faecal contents, impaired perception of rectal distension with loss of desire to defecate and rectal sphincter dysfunction. The longer the faeces remain in the colon, the drier it becomes (Larkin et al. 2018).

UNDERSTANDING THE DIGESTIVE SYSTEM THROUGH TRADITIONAL CHINESE MEDICINE: FROM THE STOMACH TO THE LARGE INTESTINE

The Stomach receives food and begins the process of digestion by breaking it down. In this activity, it is associated with the Spleen and their complementary function of digestion and collecting the essence of food in order to nourish the body. Next comes the function of the Small Intestine.

The small intestine receives and closes. The matter that comes out of it is transformed. Its function is to carry out a new digestion of the matter coming from the stomach, separating the essence from the waste. The "Clear" will be distributed throughout the body, the "Cloudy" will be concentrated in the Large Intestine. The unused water will diffuse into the bladder. If this function of separating the "Clear" from the "Cloudy" is defective, digestive anomalies will be created which will manifest themselves in the urine and faeces.

For the large intestine, its function is to receive food waste from the small intestine. From there, it absorbs excess water and expels waste. If the function of the large intestine is dysregulated, constipation or diarrhoea is observed (Auteroche and Navailh, 1992).

CHAPTER 2

SIGNS, SYMPTOMS AND DIAGNOSIS OF CONSTIPATION

Constipation is most often reported by women and people over 65. Symptoms are diverse and include infectious bowel movements, hard faeces, excessive straining, bloating and abdominal discomfort (Chey et al. 2011). Chronic constipation has been associated with considerable impairment in quality of life, can result in large individual healthcare costs and represents a burden on healthcare delivery systems (Gallegos-Orozco et al. 2012).

Pathophysiological basis of chronic constipation: The act of defecation is dependent on the coordinated functions of the colon, rectum and anus. Considering the complexity of the neuromuscular functions (sensory and motor) required to achieve planned, conscious and effective defecation. Clinically, such problems usually lead to symptoms of obstructed defecation (e.g. straining, incomplete, unsuccessful or painful evacuation, intestinal infection and abdominal pain and bloating). After excluding a multitude of secondary causes (obstructing colonic lesions, neurological, metabolic and endocrine disorders), the pathophysiology of chronic constipation can be broadly divided into problems with colonic contractile activity and therefore stool transit and pelvic floor problems (Grossi et al. 2018).

Constipation is a symptom-based syndrome defined as unsatisfactory defecation, characterised by difficulty passing stools or infrequent stools, hard stools or a feeling of incomplete evacuation. According to the Rome III criteria. Constipation is defined when two or more of the following symptoms have been present for at least three months in the last six months before diagnosis: straining >25% of bowel movements; feeling of incomplete evacuation >25% of bowel movements; less than three bowel movements per week; feeling of outlet obstruction >25% of bowel movements; manual manoeuvres to facilitate evacuation >25% of bowel movements, e.g. digital manoeuvres, supporting the pelvic floor. It should also be noted that loose stools are rarely present without the use of laxatives and the patient has insufficient criteria for irritable bowel syndrome (Schmidt et al. 2015).

According to Eriksson et al. 2008, referenced by Moreira et al. 2009, somatic, psychological and biochemical differences were demonstrated between the subtypes of irritable bowel syndrome. The subtype with a predominance of constipation scored higher on symptoms of depression and anxiety, associated with lower quality of life compared to the other subtypes.

According to Diaz and Mendez 2018, The Rome IV criteria establish functional constipation when two or more of the following are present for at least one month for infants and children up to 4 years of age. For children over four years of age, symptoms must last for at least two months:

Two or fewer bowel movements a week;

At least one episode of faecal incontinence per week after the child has acquired complete bowel control;

History of extensive faecal retention or the child's retention behaviour;

Having hard, painful faeces;

Large faecal mass on digital rectal examination;

Large in diameter faeces that cause rectal outlet obstruction.

A careful history and examination is usually sufficient to make a diagnosis. A thorough history is recommended as part of a complete assessment of a child with constipation. Key information includes the time after birth of the first bowel movement, the period during which the illness was present, the incidence of bowel movements, the consistency and size of the faeces, whether defecation is painful, whether there is blood in the faeces or on the toilet paper and whether defecation is associated with abdominal pain. It is also important to ask about contamination, which can be mistaken for diarrhoea in some parents. Medications can be associated with constipation, such as opiates, sucralfate, antacids, among others. A psychosocial history is important to assess the family structure, the number of members living at home and the relationship with them, interaction with peers and the possibility of abuse.

The physical examination should include an assessment of "alarm" signs and symptoms (in other words, fever, abdominal distension, anorexia, nausea, vomiting, weight loss or inadequate weight gain). Bloody diarrhoea in a child with constipation can be an indication of a diagnosis such as Hirschsprung's disease. On abdominal examination, distension or a palpable "mass" can be appreciated in the lower abdomen. A rectal examination should be performed to identify the presence of impacted faeces or an intra-rectal mass. Visual and digital anal inspection to ensure normal size and positioning of the anal opening and to assess for rectal prolapse.

More than half of children assessed as outpatients for abdominal pain are diagnosed with constipation. The use of X-rays in this scenario is variable: less than 5% in clinics, more than 70% in emergency services. X-rays increase the rate of diagnostic errors, remain costly and involve radiation exposure (Kearney et al. 2018).

Chronic idiopathic constipation (CIC) is one of the most common gastrointestinal disorders, with a global prevalence of 14 per cent. It is more common in women and its prevalence increases with age. There are three subtypes of CIC: dyssynergic defecation, slow transit constipation and normal transit constipation, which is the most common subtype.

Clinical assessment of the patient with constipation requires a careful anamnesis in order to identify any warning symptoms that require further investigation with colonoscopy to rule out colorectal malignancy.

Screening for hypercalcaemia, hypothyroidism and coeliac disease with appropriate blood tests should be considered.

A digital rectal examination should be carried out to assess evidence of dyssynergic defecation. If this is suspected, a more detailed investigation with high-resolution anorectal manometry should be carried out.

Anorectal biofeedback can be offered to patients with dyssynergic defecation as a way of correcting the associated impairment of the pelvic floor, abdominal wall and rectal function.

Lifestyle modifications, such as increasing dietary fibre, are the first step in managing other causes of CIC. If patients don't respond to these simple changes, treatment with osmotic laxatives and stimulants should be trialled.

Patients who do not respond to traditional laxatives should be treated with prosecretory agents such as lubiprostone, linaclotide and plecanatide, or the 5-HT4 receptor agonist prucalopride, when available.

If there is no response to pharmacological treatment, surgical intervention can be considered, but it is only suitable for a carefully selected subgroup of patients with proven slow transit constipation (Black and Ford, 2018)

PREVALENCE

The study in ten Brazilian cities revealed that: 66% reported gastrointestinal symptoms, gas symptoms 46%, abdominal distension and constipation 43%.Gastrointestinal symptoms affected the quality of life of most women (62%), especially constipation(mood (89%), concentration (88%) and sexual life (79%). Gastrointestinal symptoms are highly prevalent in Brazilian women and have a negative impact on different aspects of quality of life (mood, concentration and sexuality). The gut is an important emotional catalyst that can modulate psychological behaviour (Del'arco et al. 2017).

Chronic constipation is a very common medical problem. Depending on the definition, up to 15 per cent of the general German population report constipation symptoms, corresponding to combined prevalence rates of 14 per cent in 18 other countries Prevalence rates are typically higher for women compared to men and increase with age. In the past, constipation was often defined only by reduced stool frequency (e.g. <3 bowel movements/week). Modern definitions, such as the new Rome IV definition, recognise constipation as a polysymptomatic disorder including various aspects of disturbed defecation. A variety of primary and secondary disorders of intestinal function (motility, secretion, sensitivity) or defecation can lead to constipation. Quality of life can be substantially reduced in affected patients. On the other hand, patients may have to face preconceptions that their constipation is just a simple lifestyle problem caused by "wrong behaviour" such as inadequate nutrition, insufficient fluid intake and lack of physical activity (Andresen and Layer, 2018).

Constipation is a common gastrointestinal problem that causes a lot of expense to the community with an estimated prevalence of 1% to 80% worldwide, where the condition is characterised by a wide geographical variation. It is worth noting that the varieties of definitions have led to a wide range of prevalence (Forootan et al. 2018).

Functional constipation is a prevalent condition in childhood, around 29.6% worldwide. In the United States, it accounts for 3% to 5% of paediatric consultations and a considerable annual health care cost. Most children do not have an aetiological factor and a third continue to have problems beyond adolescence. Up to 84% of children with functional constipation suffer from faecal incontinence, while more than a third of children have primary or secondary behavioural problems due to constipation (Diaz and Mendez, 2018).

The reported prevalence of constipation in patients with advanced cancer varies between 40 and 90 per cent, more common in the population treated with opioids. Prevalence increases with age and the elderly are five times more prone to constipation than young people, due to polypharmacy, reduced mobility, reduced hydration and reduced desire to defecate. In older cancer patients receiving palliative care, constipation is one of the most prevalent symptoms, with prevalence rates ranging from 51% to 55% (Larkin et al. 2018).

Functional constipation among children (18.5%) and children and adolescents (14.1%). Quality of life was decreased in paediatric patients with functional gastrointestinal disorders. Children were more likely to qualify for a functional gastrointestinal disorder if their parents qualified for a functional gastrointestinal disorder (Robin et al. 2018).

Constipation is common in adults and children and up to 20% of the population report this symptom depending on the definitions used (2-28% of adults, 0.7-30% of children) with a higher prevalence in women. Chronic constipation, generally defined as more than 6 months of symptoms, is less common but results in 0.5 million UK GP consultations per year. In the US in 2012, primary complaint of constipation was responsible for 3.2 million medical visits resulting in direct and indirect costs of $1.7 billion. In the UK, annual spending on laxatives exceeds £80 million, with a cost of £17.4 million for prescriptions in 2012 (Grossi et al. 2018).

Constipation is one of the most common gastrointestinal symptoms in children. With an average reported prevalence of 12%, it accounts for around 25% of all paediatric gastroenterology consultations (Tambucci et al. 2017). Constipation in children is common, with a prevalence of 3-30% worldwide. Most constipation in children is functional and related to behavioural retention following an unpleasant stool event (Waterham et al. 2017).

A 44% prevalence of constipation was found in an epidemiological, cross-sectional and descriptive study of undergraduate physiotherapy students at the State University of Alagoas. The prevalence between genders was 85% for females compared to 15% for males (Bomfim et al. 2017).

In the elderly, a field study in the district of Barreiro, Sete Lagoas, Minas Gerais, showed a prevalence of 23.33% of constipation in those assessed. The authors believe that there is an association with the use of medication and the lack of physical activity in the group studied (Jesus and Diniz, 2017).

A cross-sectional study with questionnaires using the Rome III criteria in the municipality of Viçosa-MG. 140 individuals were assessed and the total prevalence was 31.4%. Women had a higher prevalence than men, with values of 32.6% and 25.6% respectively (SanfAnna *and* Ferreira, 2016).

A population-based, cross-sectional epidemiological study on bowel habits in the Brazilian population showed a prevalence of 25.2% for self-reported constipation, 37.2% for women and 10.2% among men (Schmidte et al. 2015).

Constipation negatively affects quality of life. Around 12 to 19 per cent of the population in the USA and 14 per cent in Asia suffer from symptoms of constipation (Xue et al. 2014)

Like obesity, constipation in childhood is growing and becoming one of the biggest challenges for paediatricians. Prevalence of 15.5%, faecal frequency less than or equal to 3 times a week, pain when defecating, hard, stony or clumped stools, or presence of blood. The authors indicate that there is an important relationship between constipation and obesity, and this relationship is due to the high risk of developing psychosocial problems and organic diseases (Felippi et al. 2014).

Constipation is more frequently reported by women and people over 65 years of age. It is estimated that 12-19 per cent of the general population experience chronic constipation (Chey et al. 2011).

COMMENT

Given the prevalence and severity of constipation, it is clear that health professionals are unaware of the pathophysiology of constipation and give little importance to such a serious situation. It is also clear why many patients are looking for alternative treatments to alleviate their suffering.

CONSTIPATION AND ITS RELATIONSHIP WITH OTHER DISEASES

Constipation is a recognised non-motor feature of Parkinson's and has been reported as an antecedent to diagnosis in several observational studies. A systematic review and meta-analysis shows that people with constipation are at greater risk of developing Parkinson's disease compared to those without, and that constipation can precede Parkinson's diagnosis by more than a decade (Adams-Carr et al. 2016).

Constipation appears at all stages of Parkinson's disease and increases as the disease progresses. Its impact on quality of life is no less than motor symptoms (Gan et al. 2018).

Constipation is the most common gastrointestinal symptom in patients with diabetes mellitus. Importantly, patients with constipation have a lower health-related quality of life than those without constipation. Effective therapies for constipation are limited and there is a paucity of data evaluating the treatment of constipation in diabetics (Christie et al. 2017).

Constipation is a common problem in patients with advanced cancer and a significant source of major morbidity and suffering, which is often underestimated (Larkin et al. 2018)

There is a strong association between obesity and constipation. Obesity is very common all over the world and is linked to morbidity and mortality. It has a large number of impacts on the human body. Constipation has a prevalence of 4% to 29% in various parts of the world, and is considered a major health problem, with an estimated incidence of 5% in men and 15% in women (Tantawy et al. 2017).

CONVENTIONAL TREATMENT FOR CONSTIPATION

Chronic constipation is a very common medical problem with a significant impact on patients' quality of life. Modern definitions recognise constipation as a polysymptomatic disorder, including various aspects of disturbed defecation. Current guidelines recommend a gradual approach to the management of chronic constipation. Isolated or concomitant defecation disorders should be identified and may require differentiated/additional treatment. Baseline measures include lifestyle components and bulking agents. The next step recommends treatment with conventional laxatives. In refractory patients, modern medical therapies such as the prokinetic prucalopride or the secretagogues linalotide or lubiprostone (Andresen and Layer, 2018).

Constipation is a common adverse event among opioid users with major functional and quality of life impairment High rates of opioid use for chronic non-cancer pain have been reported worldwide, despite their association with adverse events, inappropriate use and limited analgesic effect. Opioid-induced constipation is the most prevalent and disabling adverse effect associated with opioid therapy (Veiga et al. 2018).

Functional constipation is a common problem in children. Its aetiology is multifactorial, involving age, behaviour, pelvic floor function and gastrointestinal motility. Conventional treatment includes education, dietary counselling, toilet training and laxatives. However, despite this multifaceted approach, 50 per cent of children still have CF after 6 to 12 months of treatment with laxatives and 25 per cent have symptoms that persist into adulthood. As a result, functional constipation not only has a major impact on the quality of life of children and their families, but also significantly increases healthcare costs (Van Summeren et al. 2018).

According to Sharma and Rao, 2017, The initial therapeutic approach to primary constipation, regardless of aetiology, consists of dietary and lifestyle changes, such as encouraging adequate fluid and fibre intake, regular exercise and dietary modification. Laxatives are the basis of pharmacological treatment for potential long-term therapy in patients who do not respond to lifestyle or dietary modification. After a failed empirical trial of laxatives, diagnostic testing is necessary to understand the underlying anorectal and/or colonic pathophysiology. No single test provides a comprehensive assessment for primary constipation; therefore, several tests are used to provide complementary information to each other. Dyssynergic defecation, a functional defecation discrder, is an acquired behavioural disorder of defecation present in two-thirds of adult patients, where there is an inability to cocrdinate the abdominal, recto-anal and pelvic flcor muscles during attempted defecation. Biofeedback therapy is the main treatment for dyssynergic defecation, aimed at improving the coordination of the abdominal and anorectal muscles.

Constipation increases with age, decreased mobility, depression and drug use, including opiates. Other research has indicated that constipation is also associated with diabetes mellitus, hypothyroidism, diverticulitis, irritable bowel syndrome and haemorrhoids. Management of chronic constipation in the elderly is essential in order to keep the main causes of constipation under control. In some cases, associated diseases need to be treated first. Cases involving secondary causes of constipation must be eliminated if chronic constipation in the elderly is to be properly treated. Firstly, a physical examination should be carried out and the history of the illness should be traced in detail. Identifying the cause of constipation is important. In difficult cases, invasive colonoscopy may be considered. In addition, some cases of chronic constipation in the elderly are related to previous use of drugs such as opioids, beta-blockers, calcium channel blockers, diuretics and anticholinergics. Common treatments used to treat patients with chronic constipation include the use of magnesium hydroxide lubricants, polycarbophila, methylcellulose, bisacodyl and misoprostol. These are recommended by the American College of Gastroenterology. Certain cases, in which conventional treatments for chronic constipation fail, will require surgical action in the form of a total colectomy with end-to-end anastomosis and ileorectal treatment (Handaya et al. 2017).

Key points according to Rao et al. 2016. Constipation is common, impairs quality of life and consumes considerable healthcare resources Recognising whether constipation is primary or secondary is key to appropriate management A detailed history and physical examination, including digital rectal examination is important and can identify a bowel movement disorder Physiological tests such as colonic transit assessment, anorectal manometry and the balloon expulsion test can facilitate the stratification of patients with different subtypes of constipation. Newer drugs such as linaclotide and lubiprostone, laxatives and biofeedback therapy can considerably improve symptoms in patients with chronic constipation

Treatment / Management in children

Diaz and Mendez, 2018, describe: Management for constipation includes medical supervision, dietary instructions, behavioural changes and toilet training instructions. Normal fibre intake, fluid intake and level of physical activity are recommended. Non-pharmacological intervention consists of demystification, explanation and guidance for toilet training in children with a developmental age of at least 4 years. Laxatives represent the first-line treatment for constipation in childhood and, if a suitable regime is implemented, they generally bring about symptomatic improvement. Consensus guidelines recommend daily polyethylene glycol (PEG) at a dose of 1 to 1.5 g/kg per day for 3 to 6 days for initial faecal disimpaction, followed by a daily maintenance dose of 0.4 g/kg per day for at least 2 months to prevent re-accumulation. A stimulant laxative should be added if PEG alone does not cause disimpaction after 2 weeks of treatment.

Biofeedback improved defecation dynamics, but did not affect constipation. Therefore, this approach has not been supported for the management of idiopathic constipation in children Sacral nerve stimulation is a modality that has been used to treat refractory constipation by assisting extrinsic neural control of the large intestine and modulating inhibitory reflexes. This has improved the frequency of defecation in some children with functional constipation, but the effects last less than 6 months in a large group of patients. Surgical management is reserved for patients who are refractory to medical interventions. At least 10% of children with functional constipation referred to a paediatric surgeon will need an operation. The aim of surgical treatment is to produce symptom relief. Surgical options can include anal procedures, antegrade enemas, colorectal resection and intestinal diversion.

Prognosis: Most paediatric patients are treated with drug therapy and most of them improve. However, at least 30 per cent will remain symptomatic until adulthood. Factors associated with a worse prognosis are female gender, older age of onset, longer time between the presentation of symptoms and the start of therapy and longer colonic transit time (Diaz and Mendez, 2018).

Constipation, a common symptom among stroke patients that leads to increased morbidity and mortality, has been reported to be associated with an increased length of hospital stay, worse neurological prognosis and death. The incidence of constipation in patients with steatosis has been reported to range from 29 per cent to 79 per cent. In addition, patients with post-stroke constipation are more likely to experience medical complications such as pneumonia, urinary tract infection, upper gastrointestinal bleeding and recurrent stroke. Conventional drugs to manage constipation include osmotic laxatives, stimulant laxatives, secretagogues and serotonin 5-HT receptor agonists; however, side effects such as headache, diarrhoea, nausea and abdominal pain need to be considered with the use of these drugs. These side effects can have detrimental psychological and physical effects on post-stroke patients. In addition, constipation can recur after medication withdrawal, resulting in drug dependence with long-term use. Traditional Chinese Medicine (TCM) is also widely used to treat constipation. The most commonly used classical formula is Ma-Zi-Ren-Wan, while the most commonly used Chinese patent medicine is Ma-Ren-Ruan-Jiao-Nang. The most commonly used Chinese herb is Da Huang (Radixet Rhizoma Rhei). The use of alternative non-pharmacological therapies, including acupuncture, electroacupuncture and massage, has also been reported. Although the efficacy and safety of these treatments remain controversial, TCM is tolerated by patients with post-stroke constipation. As far as we know, the therapeutic effect of moxibustion in post-stenosis patients with constipation has not been well investigated. Here, we report the case of a 65-year-old man with post-stroke constipation who was successfully treated with moxibustion at the Tianshu (ST25) and Qihai (CV6) acupoints after other treatments proved ineffective (Wen et al. 2018). The findings of this case suggest that moxibustion is effective and safe for treating constipation and improving quality of life in patients with post-stroke constipation.

Rationale: Moxibustion, an important therapeutic measure in TCM, can stimulate acupuncture points to unblock meridians and collaterals, regulate qi and blood function, support health and expel pathogens. It can therefore be effective and safe for treating constipation and improving quality of life in post-stroke patients with constipation. (Wen et al. 2018)

COMMENT

When the text talks about bulking agents, it's referring to fibres. As far as treatment is concerned, it is important to realise that when pharmacological treatment is interrupted, recurrence occurs. Side effects are important, indicating the need for alternative treatment.

CENTRAL PROPOSAL OF THIS DOCUMENT

ACUPUNCTURE AND PROBIOTICS

When I studied Chinese medicine, especially acupuncture, I realised that this brilliant medicine attaches great importance to the nutrition of all the body's systems. Chinese medicine thinks of the body as a whole and not as an isolated system. If a patient chooses acupuncture to treat one symptom, many others will be treated at the same time. Like probiotics, acupuncture acts on the gut-brain axis. Acupuncture works to maintain the microbiota.

When I saw a paediatrician with symptoms of nausea and vomiting due to an intragastric balloon, I noticed that the symptoms were quickly minimised. I then commented on the finding to a professor of clinical acupuncture and mentioned the variety of acupuncture points used. He replied that I didn't need so much and indicated two points. VC 12 with bilateral CS6. The efficiency of these two points is fantastic. I tested it: On the way down a mountain in a car, my friend's father started vomiting and the frequency was constant. Worried about the progress of the journey, we decided to stop and I put on the two dots. Incredibly, the vomiting stopped. Then I thought, this is a medicine that should be available to all societies around the world. As the wise

Solomon said, a rope with three ends is harder to break. Probiotics and acupuncture are two strong ends to make the body resist the hardships of modern life.

CHAPTER 3

ACUPUNTURE

The definition of acupuncture is too simplistic, dealing only with the mechanics of needle insertion. I prefer to say that Acupuncture is, among the various parts of Chinese medicine, the most important branch for integrating the various segments of the human being. Namely: Physical, emotional, psychic and spiritual.

The state of health or illness depends on the energetic condition of the blood (xué) and energy (Qi). These two basic elements work to nourish the five organs referred to by the physiology of the 5 elements. Lung (Metal), Kidney (Water), Liver (Wood), Heart (Fire), Spleen-Pancreas (Earth). The five organs are co-responsible and each has a function for certain structures or tissues in the body. The energetic balance of the body and everything that exists is based on the relationship between the polarised and interdependent forces, Yin / Yang. And it is in this fascinating complexity that Acupuncture plays its role.

According to Liu et al. 2018, acupuncture is one of the most important components of traditional Chinese medicine and is widely used in the treatment of various diseases. In addition, its healing effect has been widely recognised and accepted worldwide by the international medical community.

USING ACUPUNCTURE FOR CONSTIPATION

Let's hear it for the scientific studies:

Anticonstipation! Medication or Acupuncture? A systematic review and meta-analysis showed that electroacupuncture was more effective than medication in improving bowel movements and overall response rate, and in reducing constipation symptoms (Zhou et al., 2017).

Acupuncture for patients with chronic functional constipation: a randomised controlled trial. A multicentre randomised clinical trial was carried out involving 684 patients with chronic functional constipation. The 684 patients were divided into 4 groups. One group with mosapride medication, and the other three groups with different acupuncture protocols. The three acupuncture treatments were as effective as mosapride in improving stool frequency and stool consistency in chronic functional constipation. However, the change in spontaneous bowel movements at week 8 was significantly lower in the mosapride group (Zheng et al., 2018).

Randomised controlled pilot study: Acupuncture showed clinically significant improvements in terms of complete spontaneous evacuation occurring more than 3 times a week and these improvements being maintained for 4 weeks after the end of treatment. The acupuncture points used were E25, E27, Bx 52 and Bx 25 (Lee et al. 2018).

Randomised controlled trial to regulate the mind and strengthen the arm in irritable bowel syndrome with a predominance of constipation. 60 patients with constipation were randomised into two groups. 30 cases were given Tianshu(E25), Zusanli(E36), Shangjuxu(E37), Taichong(F3) Sanyinjiao(BP6), Yintang, and Baihui(VG20) acupuncture, once a day, 5 treatments a week for 4 weeks. In the Western medication group, 30 cases. Oral solution was prescribed, 15 ml at a time 3 times a day for 4 weeks. Symptom and quality of life scales were adopted. The results showed that Acupuncture to Regulate the Mind and Strengthen the Spleen significantly relieves the clinical symptoms of constipation and improves patients' quality of life, and the overall efficacy is superior to oral lactulose solution and shows some long-term efficacy (Pei et al. 2015).

According to Su, 2017, two of the most common points for colds are Quchi (IG 11) and Shangjuxu (E37). IG 11 reduces body heat, not only internally, but also has applications for hot skin conditions such as fever. Due to its heat-reducing functions, it can also be used for certain types of diarrhoea and constipation. E37 is used for more acute problems such as the colon and digestive organs in general. In addition to constipation, it can be used for abdominal pain and bloating. IG11 and E37 are well suited to the treatment of irritable bowel syndrome. On a biochemical level, they increase the expression of TPH and 5-HT which are markers for healthy intestinal function. 5-HT (serotonin) which is an important marker for the health of the intestines and the general health of the body, as well as mood. These biochemical markers alone begin to explain why Traditional Chinese Medicine never separates the body from the mind and why many acupuncture points end up having such broad systemic responses.

Acupuncture has a good curative effect in the treatment of diseases of the gastrointestinal system, such as functional dyspepsia, ulcerative colitis and constipation. The regulation of gastrointestinal function by acupuncture has been found to involve many brain regions, such as the amygdala, paraventricular nucleus, locus coeruleus, raphe nucleus and dorsal nucleus of the vagus nerve (Bu et al. 2017).

A study in rats showed that acupuncture, electroacupuncture and moxibustion can all play a positive regulatory role in functional constipation in rats, with electroacupuncture having the best efficacy, followed by acupuncture (Xu et al. 2017).

Acupuncture at the points Danzhong (VC17), Qihai (VC 6), Tianshu (E25) , Neiguan (PC 6), Gongsun (BP 4), and Taichong (F3), achieves a better effect than lactulose for constipation in post-stroke patients in terms of efficacy onset, extension and long-term (Gao et al. 2017).

Combined therapy of acupuncture and herbal medicine, simple use of herbal medicine and simple use of Western medication. The combination of acupuncture and Chinese medicinal herbs improves colonic function and relieves conditions of anxiety and depression in constipated patients. The
therapeutic effects are better than those treated with the simple use of herbs and Western medication (Wang et al. 2017).

Electroacupuncture shows therapeutic effect for functional constipation. The points used were: Tianshu(E25) , Fujie (BP 14) and Shangjuxu(E37) (Lv et al. 2017).

Randomised placebo-controlled study. Auricular acupressure was applied to seven auricular acupuncture points for 10 days. The results indicated positive clinical value of auricular acupressure with magnetic beads in the treatment of constipation in the elderly. Auricular acupressure was also found to be a safe and acceptable intervention (Li et al. 2014).

Randomised controlled trial to compare acupuncture and oral medication in female functional constipation. 56 female patients were randomised into an acupuncture group and a medication group. The acupuncture points were: Zhongwan(VC12), Tianshu(E25), Guanyuan(VC4), Xiawan(VC10), Huaroumen(E24), Wailing(E26), Zhigou(TA6), Shangjuxu(E37).It was applied once a week. Lactulose as an oral medication was administered in the medication group 3 times a day. The treatment cycle was 8 weeks. The authors concluded that: Acupuncture significantly improves stool character, defecation difficulty and quality of life in females with functional constipation (Shen et al. 2017) .

Efficacy of acupuncture for the treatment of chronic constipation in childhood. 17 children constipated for 6 months were treated with five weekly placebo acupuncture sessions, followed by 10 weekly true acupuncture sessions. Panopioid activity was measured at time 0 and after 5, 10 and 15 acupuncture sessions. The frequency of bowel movements in males gradually increased compared to females and reached a maximum improvement only after 10 sessions of true acupuncture. Baseline panopioid activity was lower in constipated children compared to the control population and gradually increased to the control level after 10 true acupuncture sessions. A successful acupuncture treatment in constipated children (Broide et al. 2001).

Randomised controlled clinical trial to compare the effectiveness of electroacupuncture on functional constipation in different acupuncture prescriptions. 140 patients were randomised into 4 groups. 3 of different acupuncture prescriptions and one of Western medication, mosapride citrate. As a result, the weekly frequency of defecation is effectively increased after treatment in the three electroacupuncture groups and the efficacy is similar to mosapride citrate tablets. Quchi (IG11) and Shangjuxu (E37) significantly increase weekly defecation frequency, relieve defecation difficulty and improve quality of life. And the efficacy is sustained for 4 weeks. Thus, prescribing the acupuncture points Quchi (IG11) and Shangjuxu (E37) is the best treatment for functional constipation (Wu et al. 2014).

A meta-analysis of randomised controlled trials of acupuncture and moxibustion found that the curative rate of acupuncture and moxibustion in constipation was better than medication. The authors concluded that acupuncture exerts an advantage when compared to routine medication treatment (Du et al. 2012)

Acupuncture increases the amplitude of jejunal motility in constipated rats. Acupuncture points IG 11 and E 37 (Qin et al. 2014).

The clinical efficacy of acupuncture and moxibustion for post-stroke constipation was systematically reviewed and the results indicated that acupuncture was effective for post-stroke constipation and has advantages compared to other therapies (Yang et al. 2014).

Irritable bowel syndrome is a functional disorder of the intestine that causes recurrent abdominal pain. Epidemiological data shows that the incidence rate of irritable bowel syndrome is very high at around 25%. Most drugs can lead to addiction and toxic side effects. Moxibustion is an important component of traditional Chinese medicine and has been used to treat the abdominal pain of irritable bowel syndrome for several thousand years in China. As a gentle treatment, moxibustion has been widely applied in the clinical treatment of visceral pain. In recent years, it has played an irreplaceable role in alternative medicine. Extensive clinical studies have shown that moxibustion for the treatment of visceral pain is simple, convenient and inexpensive and is being accepted by an increasing number of patients (Huang et al. 2014).

DOCUMENTED EFFECTS OF ACUPUNCTURE

The term "acupuncture" refers to needles, moxibustion, acupressure, laser acupuncture, electric acupuncture, microsystem acupuncture such as ear, face, hand and scalp acupuncture. Of all the complementary medical systems, acupuncture enjoys the most credibility (Yeh et al. 2017).

According to Liu et al. 2018, acupuncture is one of the most important components of traditional Chinese medicine and is widely used in the treatment of various diseases. In addition, its healing effect has been widely recognised and accepted worldwide by the international medical community. Acupuncture through Baihui (DU20) to Qubin (GB7) has a "reparative" function, inducing the expression of neurotrophic factor derived from the endogenous glial cell lineage during acute cerebral haemorrhage. More specifically, acupuncture can improve the recovery of neural stem cells. Acupuncture can also antagonise the inflammatory brain injury generated by cerebral haemorrhage. Thus, acupuncture is an effective means of reducing the expression of inflammatory mediators in the nervous system. In addition, according to current reports, acupuncture has greatly contributed to reducing the rate of stroke-induced disability and improving the recovery of neural function. Studies investigating acupuncture have found it to be a highly potent therapy for reducing neural inflammation, suppressing cell apoptosis and alleviating nerve dysfunction after stroke. Taken together, the studies show that acupuncture has great potential for treating cerebral haemorrhage by inhibiting inflammation.

Acupuncture is increasingly used as a therapy in the United States. Manual acupuncture and electroacupuncture have been used to treat a variety of illnesses in the Far East for centuries (Zhou and Benharash, 2014).

The results of clinical trials indicate that acupuncture has a better effect for slow transit constipation, improving severity, increasing defecation frequency, reducing abdominal distension, easing patients' psychological discomfort and increasing activity of daily living (Yu and Xu, 2016).

Lactobacillus and Bifidobacterium increased after acupuncture. The results show that acupuncture can play an ecological role in the gastrointestinal tract, activating gastrointestinal function by adjusting the body's immune system. By restoring the intestinal microbial balance, acupuncture has been shown to be effective in treating chronic constipation, chronic diarrhoea, acute bacillary dysentery and irritable bowel syndrome (Xu et al. 2013). Dysbiosis can contribute to constipation and its symptoms (Riezzo et al. 2018). Chronic constipation is a prevalent functional gastrointestinal disease accompanied by intestinal dysbiosis (Cao et al. 2017).

A bibliometric study on clinical indications for acupuncture indicates that: The increase in clinical indications reflects the extensive application of acupuncture and can help provide a better service for people's health (He et al. 2012).

For at least 2,500 years, acupuncture has been an integral part of traditional Chinese medicine. However, recently, as more people in Western countries are diagnosed with chronic illnesses that are poorly treated with modern medical therapies, many are turning to acupuncture and other forms of alternative medical treatments. Based on the theory of harmonious floating qi being the basis of good health, acupuncture focuses on restoring qi by manipulating the complementary and opposing elements of yin and yang (Vanderploeg and Yi, 2009).

Auricular acupressure was effective in relieving constipation in breast cancer patients receiving chemotherapy. A randomised control trial (Shin and Park, 2018)

MECHANISMS AND REASONS FOR THE BENEFICIAL EFFECTS OF ACUPUNCTURE

Electro-acupuncture at specific acupoints can improve intestinal motility in constipation by altering the enteric nervous system and differentially affecting excitatory and inhibitory neurons, restoring coordination between contraction and relaxation muscles, and working together with the central nervous system and peripheral neural pathways (Liang et al. 2018).

Effects of acupuncture on various systems and symptoms: It has been determined that the levels of endophin-1, beta endorphin, enkephalin, and serotonin increase in plasma and brain tissue through the application of acupuncture. Endophin-1, beta endorphin, enkephalin, serotonin and dopamine cause analgesia, sedation and recovery in motor functions. It also has immunomodulatory effects on the immune system and lipolytic effects on metabolism. It is therefore used to treat pain syndromes such as migraine, fibromyalgia, osteoarthritis and trigeminal neuralgia. Gastrointestinal disorders, such as gastrointestinal motility disorders and gastritis, psychological illnesses such as depression, anxiety and panic attacks, and in the rehabilitation of hemiplegia and obesity (Cabyoglu et al. 2016).

Acupuncture has been practised in China for over 2000 years to treat a variety of diseases based on the "meridian theory", as described in the Yellow Emperor's Classic of Internal Medicine. The meridians involved with gastrointestinal processes are (E) 36, 37, which overlap with the deep peroneal nerve. Acupuncture needles, manipulated manually or stimulated with low current and frequency, have been documentec as the neurophysiological basis for modulating the activity of peripheral and central neural pathways (Zhou and Benharash, 2014).

Acupuncture and the use of meridians have long been used to treat a variety of pathological conditions. Recently, discovered acupuncture mechanisms have revealed that acupuncture points are rich in sensory nerve receptors. The points become painful because their sensory nerve receptors are pathologically sensitised. This sensitised condition is a dynamic process that is triggered by altered homeostasis and returns to baseline after the insult is corrected by acupuncture. Although the meridians were created on the basis of empirical observations, studies continue to shed light on the complex neurophysiological mechanisms behind acupuncture. Although we may be many years away from fully understanding the neurophysiological mechanisms of acupuncture points, we should keep the meridian theory in mind (Zhou and Benharash, 2014).

Acupuncture has been used to treat the symptoms of many conditions. Acupuncture does not have a single form of action, but rather a range of effects on various functions. The main therapeutic effects of acupuncture are the stimulation of the nervous system to release various substances such as opioid peptides and serotonin (White, 2009).

Acupuncture modulates various biomechanical responses, such as prokinetic, antiemetic and antinociceptive effects. Acupuncture treatment involves the insertion of fine needles into the skin and underlying muscle and the needles are stimulated manually or electrically. Acupuncture thus stimulates the somatic afferent nerves of the skin and muscles. The somatic sensory information from the body is carried to the cortex area of the brain. Somatic sensory fibres also project to the various nuclei, including the brainstem, periaqueductal grey (PAG) and the paraventricular nucleus (PVN) of the hypothalamus. Somatosensory pathways stimulated by acupuncture activate these nuclei. Brainstem activation modulates the imbalance between sympathetic and parasympathetic activity. The opioid released from the PAG is involved in mediating the antiemetic and antinociceptive effects of acupuncture. The release of oxytocin from the PVN mediates the anti-stress and antinociceptive effects of acupuncture. Acupuncture may be effective in patients with functional gastrointestinal (GI) disorders because of its effects on gastrointestinal motility and visceral pain (Takahashi, 2013).

SUMMARY FOR ACUPUNCTURE

Due to the evidence of the therapeutic effects of acupuncture in the treatment of various pathologies, I realised during the searches of existing studies that more studies are underway to better clarify the mechanisms of action of acupuncture.

For the proposal of acupuncture in the treatment of constipation, the studies answer our question and make it clear that acupuncture is a safe and effective therapy for relieving the suffering resulting from constipation.

The studies presented a variety of acupuncture points. However, two points were shown to be very effective. E37 and IG11.

USE OF PROBIOTICS IN CONSTIPATION

If the intestines are the gateway to health and disease, they need to be colonised by agents of defence and multiple actions involving digestion, absorption and nutrition. Probiotics, popularly known as good bacteria, play a magnificent role in the skin, mouth and the entire digestive and absorption system, the male and female urogenital systems, as well as maintaining a direct relationship with the brain.

I first learned about probiotics in 1996, when a researcher presented his experiments at a paediatric nutrition congress in São Paulo, Brazil. I was amazed by the role of probiotics. I started to advise, and by 1998, when I finished my degree in nutrition, probiotics were a priority in my prescriptions. In practice, the evidence made me increasingly enthusiastic as many of the patients' symptoms disappeared. So I prescribed them for muscle hypertrophy in order to increase nutrient absorption, and for weight loss, since most patients suffer from constipation. For constipation in children, adolescents, adults and the elderly. For diabetes, fatty liver and all problems involving dysbiosis. When I realised that health professionals and society were not aware of the existence of probiotics, I started to spread the word. In 2009, we wrote the article Probiotics and Physical Exercise in Constipation. In the research for the article, we realised that constipation is a serious situation and one that is of little importance to the medical community. Since constipation continues to cause so much suffering, I'm reinforcing the importance of probiotics with this prescription.

DEFINITION OF PROBIOTICS AND THEIR ROLE IN HEALTH AND DISEASE

Probiotics are defined by the Food and Drug Administration (FDA) and the WHO as "live microorganisms that, when administered in adequate amounts, confer a health benefit on the host". (Meng et al. 2018)

The human gut is perhaps one of the most complex networks in the body and is colonised by trillions of microorganisms, including bacteria, archaea, fungi, protists and viruses, among which bacteria are the main inhabitants. The gut microbiota are closely and functionally related to humans and play an important and unique role in human health and disease (Meng et al. 2018).

The human intestinal microbiota consists of trillions of microbes that form a complex ecosystem. Protection against microbial invasion is provided by the intestinal barrier. The intestinal barrier has multiple lines of defence, including commensal bacteria, which competitively inhibit the colonisation of pathogenic bacteria and the production of metabolically protective compounds such as butyrate. Impaired intestinal barrier function can result in a local or systemic immune response, mast cell degranulation, neuroinflammation and activation of the afferent vagus nerve. In addition, commensal bacterial species, such as Lactobacillus plantarum, regulate intestinal epithelial integrity by stimulating the Toll-like receptor 2. The gut microbiota changes dramatically during pregnancy and intrinsic factors (such as stress) as well as extrinsic factors (such as diet and drugs) influence the composition and activity of the gut microbiome throughout life. Microbial metabolites and short-chain fatty acids affect gut brain signalling and the immune response. The gut microbiota plays a regulatory role in anxiety, mood, cognition and pain, which is exerted via the gut-brain axis. The intake of prebiotics or probiotics has been used to treat a number of conditions, including constipation, allergic reactions and infections in childhood, and irritable bowel syndrome. The gut microbiome affects virtually every aspect of human health (Mohajeri et al. 2018).

Lifestyle and dietary changes cause microbiome instability, producing significant disturbances in the gut microbiome. Analysis of the intestinal microbiome revealed that the use of proton pump inhibitors (PPIs) was associated with a significant decrease in the diversity of the intestinal microbiota and with significant changes of around 20 per cent in the bacterial rate. This adverse effect of PPIs on bacterial diversity was greater than for any other class of drugs, including antibiotics. PPIs depleted beneficial bacteria, such as the Ruminococcaceae family and Bifidobacterium, and increased potentially harmful bacteria, including the genera Enterococcus, Streptococcus, Staphylococcus and Escherichia coli. The results suggested that PPIs lowered the gastric acid barrier, as the species found in the oral microbiome of PPI users were more abundant in the gut than in non-users. It is increasingly observed that the use of PPIs is associated with an increase in the incidence of enteric infections such as Clostridium difficile and Campylobacter (Mohajeri et al. 2018).

According to Mohajeri et al. 2018, intestinal microbiota imbalance, or dysbiosis, is thought to play a significant role in the pathogenesis of intestinal disorders, and extra-intestinal disorders, including allergies, asthma, type 1 diabetes, cardiovascular disease, metabolic syndrome and obesity. Chemotherapy-induced mucositis, which occurs in the mouth and intestine, results from damage to the mucosal barrier and can result in bacteraemia, which is the abnormal presence of bacteria in the blood. It has been suggested that commensal intestinal bacteria may play a key role in ameliorating inflammation and bacteraemia. In a mouse model of chemotherapy-induced mucositis, the number and diversity of the faecal microbiota was substantially decreased, including anaerobes and streptococci, although there was a relative increase in Bacteroides. Supporting the beneficial anaerobic microbiota during chemotherapy could therefore improve the treatment and quality of life of cancer patients.

The development of the perinatal intestinal microbiota is influenced by multiple factors, including gestational age, type of delivery, maternal microbiota, infant feeding method, genetics and environmental factors such as food choice. Microbial diversity increases dramatically during the first months of infancy. At birth, the microbiota is aerobic, with low numbers and low diversity, with the most common bacteria being facultative anaerobes and members of the Enterobacteriaceae phylum. Within a few days, the intestinal environment becomes anaerobic, resulting in the growth of bacteria such as Bifidobacterium, which is the dominant genus of bacteria in the baby's gut in the first few months of life. With the introduction of solid foods, a more adult microbiome begins to develop from the age of 6 months, dominated by Firmicutes and Bacteriodetes. Development of the intestinal microbiome during childhood. The development of the infant microbiome depends on various factors, such as the method of infant feeding, the diet and the environment. In addition, the mode of delivery (vaginal or caesarean section) affects the early microbiome. The transfer of bacteria from mother to foetus has also been demonstrated, indicating that pregnancy may be important for the colonisation of the foetal/infant gut (Mohajeri et al. 2018).

Factors that promote a healthy microbiota in neonates include vaginal delivery, full-term birth, breastfeeding and exposure to a variety of microorganisms. In contrast, a caesarean section, premature birth, formula milk and exposure to antibiotics have a negative impact on the diversity and composition of the microbiota in babies. Premature newborns show a delayed colonisation of the intestinal microbiota with Bifidobacterium, and have a high prevalence of Enterobacteriaceae, Staphylococcus and Enterococcaceae. Newborns born vaginally have an increased prevalence of maternal microbiota derived from the vagina and intestine (e.g. Lactobacillus, Prevotella and Sneathia) compared to neonates born by caesarean section. Children born by caesarean section have a relatively high prevalence of skin bacteria, such as Staphylococcus, Propionibacterium and Corynebacterium, compared to those delivered vaginally. Maternal antibiotic treatment that results in reduced use of human milk and prolonged hospitalisation usually causes an increase in the prevalence of

Proteobacteria, Firmicutes, Enterobacteriaceae (E. coli and Klebsiella spp.), Staphylococcus, Propionibacterium and Corynebacterium. Powdered milk is associated with increased bacterial diversity, increased prevalence of Bacteroides fragilis, Clostridium difficile and E. coli and decreased prevalence of bifidobacteria (Mohajeri et al. 2018).

The development of the neonatal microbiome depends on several factors. It has long been known that the mode of birth, the mode of feeding and exposure to antibiotics affect the development of the neonatal microbiome. Since treatment with pre- or probiotics can also affect the neonatal microbiome, such treatments can be effective options for optimising the development of the neonatal microbiome. It has become clear that the foetus and placenta are not sterile and the transfer of bacteria occurs from mother to foetus during pregnancy (Mohajeri et al. 2018).

Microbiome and the gut-brain axis. Bidirectional signalling between the gut microbiota, the gut and the brain occurs via neuronal pathways involving the central and enteric nervous systems, as well as the circulatory system. The latter includes the involvement of the hypothalamic-pituitary-adrenal (HPA) axis, immune system regulators, hormones, bacterial metabolites such as AGCCs and neurotransmitters. Pre-clinical studies have shown effects of gut microbiota on nociceptive reflexes, feeding, emotional and social behaviour, the stress response and brain neurochemistry. The gut microbiota is essential for normal social development in the mouse and has been implicated in neurodevelopmental disorders, including autism spectrum disorder. Germ-free mice have an exaggerated response to stress compared to control animals. These mice also exhibit increased motor activity and less anxiety-like behaviour compared to control mice. Administration of the probiotic L. rhamnosus (JB-1) to mice reduced stress-induced corticosterone levels and anxiety-related behaviour. These data strongly emphasise the importance of the microbiome-intestine-brain axis for normal neurological development and function. (Mohajeri et al. 2018)

Probiotics are "live microorganisms which, when administered in adequate quantities, confer a health benefit on the host". The main benefit of probiotics is their contribution to maintaining a balanced microbiota and therefore creating a favourable intestinal environment. In addition, probiotics support the health of the digestive tract and the immune system. The positive effects of probiotics on intestinal health in a variety of conditions have been evaluated by randomised controlled clinical trials (Agamennone et al. 2018).

Yoon et al. 2018 demonstrated that probiotics improve stool consistency in patients with chronic constipation: a randomised, double-blind, placebo-controlled study. Probiotics significantly improved stool consistency in patients with chronic constipation. In addition, the beneficial effect of L. plantarum on stool consistency remained after probiotic supplementation was discontinued.

Chronic functional constipation is a common type of intestinal disease that occurs in children, adults and the elderly. This disease not only has a major influence on physiological function, but also results in varying degrees of psychological barriers. Currently, constipation treatments continue to rely on traditional methods such as purgative therapy and surgery. However, these approaches can impair intestinal function. Recent research into intestinal diseases and intestinal microbiota has gradually revealed a connection between constipation and disturbance of the intestinal flora, providing a theoretical basis for microbial treatment in chronic constipation. Microbial treatment mainly includes probiotic preparations such as probiotics, prebiotics, symbiotics and faecal microbiota transplantation (FMT). Due to their safety, convenience and curative effect, probiotic preparations have been widely accepted, especially the gradually developed FMT with greater curative effects. Microbial treatment improves clinical symptoms, promotes the recovery of intestinal flora and does not present complications during the treatment process. Compared to traditional treatments, microbial treatment in chronic constipation has advantages and is worthy of further promotion, from clinical research to clinical application (Huang et al. 2018).

Chronic constipation is often accompanied by emotional disorders such as depression and anxiety. Study in mice. The aim of the study was to determine whether the administration of a multi-species probiotic can decrease depressive behaviours via the gut-brain axis. In conclusion. Probiotics can alleviate constipation-induced depression by protecting neuronal health (Xu et al. 2018).

The composition of the intestinal microbiota is very important in human health. Gastrointestinal disorders are among the symptoms commonly reported by individuals diagnosed with chronic diseases such as inflammatory bowel disease, autism and chronic fatigue syndrome. The effects of probiotics and prebiotics for dysbiosis have been reported in many studies. Intestinal nosodes made from intestinal bacteria from European patients were administered to Japanese patients suffering from gastrointestinal disorders such as constipation and diarrhoea to determine their therapeutic efficacy. Of the 23 patients analysed, 69.6% showed some kind of improvement, and no deleterious effects of taking intestinal nosodes were observed; 26% of patients showed significant improvement or were "cured" (Uchiyama-Tanaka, 2018).

Analysis of randomised, placebo-controlled studies suggests that the administration of probiotics significantly improved constipation in elderly individuals (Martinez-Martinez et al. 2017).

The short-term efficacy of faecal microbiota transplantation combined with soluble dietary fibre and probiotics in the treatment of slow transit constipation: The combination of faecal microbiota transplantation with soluble dietary fibre and probiotics is safe and effective in the treatment of slow transit constipation, which can significantly improve symptoms and quality of life (Ge et al. 2016).

Controlled study in children aged 4 to 16 with irritable bowel syndrome: Administration of symbiotics and probiotics resulted in significant improvements in initial complaints when compared to prebiotics. In addition, there were significantly more patients with full recovery of irritable bowel syndrome symptoms in the symbiotic group than in the prebiotic group (Basturk and co. 2016).

A randomised, double-blind, placebo-controlled study: The efficacy of multispecies probiotic supplementation in relieving the symptoms of irritable bowel syndrome associated with constipation. Multispecies probiotic supplementation is effective in individuals with irritable bowel syndrome with a predominance of constipation, and induces a different assessment in the composition of the intestinal microbiota (Mezzasalma et al. 2016).

A pilot study shows that kefir has positive effects on constipation symptoms. Results also suggest that kefir improves intestinal satisfaction and speeds up colonic transit. Kefir, a probiotic fermented milk product (Turan et al. 2014).

The Mexican consensus on probiotics in gastroenterology.

The consensus group recommends the use of probiotics under the following clinical conditions: prevention of antibiotic-associated diarrhoea, treatment of acute infectious diarrhoea, prevention of Clostridium difficile infection and necrotizing enterocolitis, reduction of adverse events of Helicobacter pylori eradication therapy, relief of symptoms of irritable bowel syndrome, the treatment of functional constipation in adults, and the induction and maintenance of remission in patients with ulcerative colitis and pouchitis, and the treatment of covert and overt encephalopathy (Valdovinos et al. 2017).

To corroborate and compare with the current references in this document, I'm sharing some of the references from our 2009 article (Moreira et al. 2009).

[A Persian version of the Old Testament (Genesis 18:8) states that Abraham owed his long life to drinking sour milk (Schrezenmeir and Vrese, 2001).

There is good evidence that the complex microbial flora present in the gastrointestinal tract of all warm-blooded animals is effective in providing resistance to disease. However, the composition of this protective flora can be altered by environmental and dietary influences, making the host animal susceptible to disease and or reducing its efficiency of food utilisation (Fuller, 1989).

The microbiota of the human gastrointestinal tract plays a fundamental role in nutrition and health. Through the fermentation process, intestinal bacteria metabolise different substrates into end products such as short-chain fatty acids and gases. However, under certain circumstances, the fermentation process can produce undesirable metabolites. This can lead to acute and chronic diseases. In addition, the intestinal flora can become contaminated by transient pathogens that can disrupt the normal structure of the community and lead to inflammatory bowel disease and colonic cancer. It is therefore important that the intestinal microflora is controlled and sustained in an optimal way. Many different environmental factors can affect the ecology of the gut microbiota; these include diet, medication, stress, age and general living conditions. Knowledge of the gut microbiota and their interactions has led to the development of dietary strategies that serve to maintain or even improve normal gastrointestinal microbiology (Gibson and Fuller, 2000).

Our intestinal flora comprises more than 100 trillion bacteria that play an important role in maintaining good health. Probiotics provide a barrier against pathogenic bacteria by aiding digestion, promoting assimilation and absorption of food and building immune response by increasing the host's ability to resist infection (Agarwal, 2008).

Probiotics can be considered the first functional food. They are non-pathogenic organisms that retain viability during storage and survive passage through the stomach and small intestine. Dairy bacteria can improve intestinal motility and relieve constipation, especially in the elderly (Otles et al. 2003).

Probiotics are administered in situations of intestinal disorders caused by specific factors such as inadequate diets, non-antibiotic and antibiotic drugs, clinical imbalances, surgery of the digestive system and in any situation that causes a disruption in the flora of the gastrointestinal tract that makes the host susceptible to disease (Karkow et al. 2007).

Drouault-Holowacz et al. 2008, in their double-blind randomised controlled study, investigated the efficacy of a combination of probiotics in 100 patients with irritable bowel syndrome (IBS) for four weeks. They were supplemented once a day with four strains of dairy bacteria. The symptoms monitored were discomfort, abdominal pain, stool frequency and quality of life. The combination of probiotics was superior to placebo in relieving IBS symptoms. Interesting results were observed such as a significant reduction in abdominal pain, a significant increase in the number of bowel movements in constipated patients and patients reported a significant improvement in flatulence. The authors report that the mechanisms for these effects are still poorly understood and need to be studied further. Evaluation is difficult and there is no gold standard. For quality of life, despite the short duration of supplementation, the data showed positive and significant improvements in quality of life from the start of probiotic supplementation to the end in specific items relating to discomfort and digestive disorders, and especially flatulence and bloating.

Constipation is a common problem, and usually refers to less than 2-3 bowel movements per week, with a small, dry volume accompanied by difficulty defecating. 135 Chinese women aged 25-65 with a diagnosis of constipation were studied for a fortnight with a probiotic containing fermented milk, and the results show that the fermented product was well tolerated by all patients and had no reported adverse effects and had a significant positive effect by increasing stool frequency by 40% and 58% after 1ª and 2ª week respectively. The authors conclude that the study showed a beneficial effect of fermented milk containing Bifidobactérium lactis on improving stool frequency, consistency and defecation in women with constipation (Yang et al. 2008).

A pilot study carried out by Bekkali et al. in 2007 showed that a mixture of probiotics (Bifidobactérium and Lactobacillus) increased the frequency of bowel movements in constipated children with a defecation frequency of less than 3 times a week. The probiotic mixture was also effective for faecal incontinence and in reducing the presence of abdominal pain. And no side effects were found in this study. The authors conclude from the results that a mixture of bifidobacterium and lactobacillus produce lactic, acetic and other acids, resulting in a reduction of ph in the colon that are effective in increasing colonic motility, subsequently leading to a decrease in colonic transit time.

For three months, 129 patients, including 38 men and 91 women with an average age of 44, were treated in a multicentre study with a symbiotic preparation (probiotic and prebiotic) in constipation-predominant IBS. The treatment was well tolerated and there were no side effects, and it showed a strong positive effect on stool frequency and a significant reduction in pain and bloating symptoms between the 1st and 3rd month of treatment (Dughera et al. 2007).

COMMENT

Comparing the studies we presented in 2009 with the current ones, the level of evidence for probiotics continues to increase and doubts about their mechanisms of action are diminishing. Probiotics are therefore safe and effective for the treatment of intestinal syndromes.

DOCUMENTED EFFECTS OF PROBIOTICS

The intestinal microbiota promotes healthy effects in the host and prevents diseases. Probiotics (probios, for life) are defined as "live microorganisms that, when administered in adequate quantities, confer a health benefit on the host". In the early 1900s, Louis Pasteur identified the microorganisms responsible for the fermentation process, while E. Metchnikoff linked the greater longevity of rural Bulgarian peoples to the regular consumption of fermented dairy products such as yoghurt. He suggested that lactobacilli could neutralise the putrefactive effects of gastrointestinal metabolism that contributed to disease and ageing. Hippocrates declared 2000 years earlier that "death is in the gut". Metchnikoff considered lactobacilli to be probiotics ("probios", conducive to the life of the host as opposed to antibiotics); probiotics can have a positive influence on health and prevent ageing. During the Neolithic period of the Stone Age, the domestication of animals took place and man began to obtain fermented food. Probably random counts in favourable environments played an important role. Faecal microbiota transplantation dates back to a 4th century Chinese manual for food poisoning or severe diarrhoea. To this day, faecal transplantation cures Clostridium difficile infections more effectively than vancomycin and prevents recurrence (Gasbarrini et al. 2016).

A randomised, placebo-controlled clinical trial indicated that probiotic supplementation had beneficial effects on glycaemic control and cardiometabolic risk markers (Mafi et al. 2018).

The effects of probiotics on total cholesterol: A meta-analysis of randomised clinical trials. Thirty-two studies including 1971 patients met the inclusion criteria. The results of this analysis showed that, compared to the control group, total serum cholesterol was significantly reduced in the probiotic group. Probiotics in capsule form may contribute to a better healing effect (Wang et al. 2018).

Dysbiosis of the gut microbiota in the elderly can cause a permeable gut, which can result in silent systemic inflammation and promote neuroinflammation - a relevant pathomechanism in the onset of Alzheimer's disease. Rebalancing the microbiome can have a beneficial impact on gut inflammation and immune activation. Supplementing Alzheimer's disease patients with a multispecies probiotic influences the composition of gut bacteria as well as the metabolism of tryptophan in serum (Leblhuber et al. 2018).

A randomised controlled study. There was a significant trend towards a reduction in the prevalence of hyperglycaemia in the probiotic and symbiotic groups, as well as in hypertension in the probiotic group. The decreases in the prevalence of metabolic syndrome were significant after taking probiotic and symbiotic supplementation compared to the control group. In conclusion, the authors describe that the potential benefits of probiotic and symbiotic use for the treatment of metabolic syndrome in pre-diabetes was supported by the results of the present study, which may provide an important strategy to combat the diseases associated with metabolic syndrome (Kassaian et al. 2018).

Oral probiotics had a significant effect on preventing preterm birth before 32 weeks of gestation. The probiotic group showed longer gestation, higher birth weight, lower rates of chorioamnionitis and higher rates of normal vaginal flora compared to the non-probiotic group (Kirihara et al. 2018).

The administration of probiotics for the prevention and treatment of a variety of paediatric infectious diseases has received increasing attention worldwide. Many scientific reports from different societies, such as the European Society for Paediatric Gastroenterology, Hepatology and Nutrition (ESPGHAN), the American Academy of Paediatrics, the World Gastroenterology Organisation and the Canadian Paediatric Society, have indicated the benefits of probiotics, supporting recommendations for the use of probiotics in the treatment of acute gastroenteritis and the reduction of antibiotic-associated diarrhoea (AAD). Several important mechanisms underpinning the observed beneficial effects of probiotics include secretion of antimicrobial substances, competitive adherence to the mucosa and epithelium, strengthening of the intestinal epithelial barrier and modulation of the immune system. Probiotic effects that are mediated by these mechanisms are an important question that needs to be addressed. In addition, pro-inflammatory transcription factors, cytokines and apoptosis-related enzymes can also be affected by probiotic strains (Plaza-Diaz et al. 2018).

Constipation is a common symptom that affects up to 30 per cent of the Western population and is strongly associated with the presence of intestinal methanogens, which can directly inhibit motor activity. The authors emphasise that Lactobacillus reuteri has a beneficial effect on chronic constipation through a significant decrease in methane (CH4) production (Ojetti et al. 2017).

Dysbiosis (a microbial imbalance) is linked to disease and bone loss; however, more importantly, the reverse is also true: treatment with probiotics can beneficially modulate the gut microbiota to improve health, including bone health. Taken together, ancient texts and recent data indicate the potential of probiotics to maintain bone health throughout life (Schepper et al. 2017).

Lactobacilli can be especially useful for women with a history of urinary tract infections, recurrent and complicated by prolonged use of antibiotics. Probiotics do not cause antibiotic resistance and can offer other health benefits due to vaginal recolonisation (Gupta et al. 2017).

According to Scourboutakos et al. 2017, considering the wide range of diseases and health conditions for which probiotics have been shown to have benefits, many of the dosages contained in probiotic food products are too low to provide the benefits demonstrated in clinical trials.

Two meta-analyses showed that the consumption of probiotics (daily intake of 107 to 1010 CFU in any form for up to 3 months) significantly reduced the duration, frequency and use of antibiotics, and absenteeism from work (Lenoir-Wiikoop et al. 2016).

The use of probiotics dates back to a time before microbes were discovered.

Fermented dairy products were depicted in Egyptian hieroglyphics, and fermented yak milk has traditionally been used by Tibetan nomads to preserve milk during their long treks. In 1906, Henry Tissier isolated Bifidobacterium from an infant and claimed that it could displace pathogenic bacteria in the intestine. These discoveries helped catalyse research into health-promoting microbes and their role in disease prevention. One of the first human studies, in 1922, used Lactobacillus acidophilus on 30 patients with chronic constipation, diarrhoea or eczema and found improvements for all three conditions. It wasn't until 10 years later, in 1932, that a study confirmed the effect of L. acidophilus on patients with constipation and mental illness (MacFarland, 2015).

CHAPTER 4

MECHANISMS AND REASONS FOR THE BENEFICIAL EFFECTS OF PROBIOTICS

The paediatric population is continually at risk of developing infectious and inflammatory diseases. Treatment for infections, particularly gastrointestinal conditions, centres on oral or intravenous rehydration, nutritional support and, in certain cases, antibiotics. Over the last decade, the administration of probiotics and symbiotics for the prevention and treatment of different acute and chronic infectious diseases has increased considerably. Probiotic microorganisms are mainly used as treatments because they can stimulate changes in the intestinal microbial ecosystem and improve the host's immune status. The beneficial impact of probiotics is mediated by different mechanisms. These mechanisms include the ability of probiotics to increase intestinal barrier function, prevent bacterial transfer and modulate inflammation by signalling the immune receptor cascade, as well as their ability to regulate the expression of selected intestinal genes in the host. However, with regard to paediatric intestinal diseases, information regarding these key mechanisms of action is scarce, particularly for immune-mediated mechanisms of action (Plaza-Diaz et al. 2018).

The human intestinal microflora plays several beneficial roles in the host, including maintaining the host's immune homeostasis. Probiotics modulate the intestinal bacterial community, alter the intestinal lumen and favour an anti-inflammatory environment, which results in improved intestinal barrier integrity and reduced bacterial translocation and its cellular components, leading to liver protection. To date, no clear mechanisms have been identified for the beneficial actions of probiotics. Furthermore, the beneficial effects of probiotics are specific to the strains and their functional properties, so the selection of potent and active probiotic strains is very important to achieve the target therapeutic effect. The results of this study increase our understanding of the molecular mechanisms by which lactobacillus strains decrease hepatic steatosis and attenuate the LPS-induced inflammatory response in both HepG2 and TPH-1 cells. Collectively, the data presented here indicate that lactobacillus exert profound immunoregulatory effects on liver cells that are hypo-responsive to LPS. Lipopolysaccharide (LPS) is a major cellular component of Gram-negative bacteria, which is capable of inducing pro-inflammatory pathways that lead to inflammation and progression of various diseases, including alcoholic and non-alcoholic liver diseases. During intestinal barrier disruption, LPS migrates from the intestinal lumen to other extra-intestinal body organs, including the liver via the portal vein, where it activates hepatic and extra-hepatic macrophages to produce reactive oxygen species and pro-inflammatory cytokines/chemokines such as IL -1, IL-6, IL-8 and TNF-a, resulting in liver inflammation (Kanmani and Kim, 2018).

An important advance in recent years has been the recognition that there are close links between the gut microbiota and the host's metabolism, and that the microbiota is an important environmental factor contributing to obesity and its complications, such as insulin resistance, type 2 diabetes, cardiovascular disease and non-alcoholic fatty liver disease (NAFLD). In fact, it has been shown that intestinal bacteria can participate in the digestion of otherwise indigestible dietary polysaccharides, thus influencing the amount of calories extracted from food. In addition, consumption of a high-fat diet is associated with loosening of intestinal junctions, increased intestinal permeability and elevated systemic levels of lipopolysaccharides (LPS). This "metabolic endotoxaemia" determines low-grade systemic inflammation, affecting insulin signalling and contributing to insulin resistance and its complications. Different pathogenic pathways are involved in the development of NAFLD and, in recent years, growing evidence has indicated the involvement of intestinal dysbiosis. While we await further research to understand when and how we should better modulate the gut microbiota in NAFLD patients, the very favourable risk profile already allows us to 'test' probiotics/prebiotics, since the currently recommended therapeutic measures have failed (Vespasiani-Gentilucci et al. 2018).

The intestinal microbiota influences the health of the host, especially with regard to intestinal immune homeostasis and the intestinal immune response. Recent findings have highlighted that changes in the microbiota modulate the host's immune system by modulating tryptophan metabolism (Gão et al, 2018).

The intestinal microbiota maintains a complex mutual interaction with different host organs. While under normal conditions this natural community of trillions of microorganisms contributes greatly to human health, intestinal dysbiosis is linked to the onset or worsening of various chronic systemic diseases. Recent studies show that the mechanisms of action of kefir (symbiotic) in cardiometabolic diseases include the recruitment of endothelial progenitor cells, improvement of the vagal/sympathetic nervous system balance, reduction of excessive generation of reactive oxygen species, inhibition of angiotensin-converting enzyme, anti-inflammatory cytokine profile and alteration of the intestinal microbiota. The intestinal microbiota refers to a broad community of non-pathogenic microorganisms that cohabit with enterocytes symbiotically, providing a variety of effects in terms of benefits for the host, creating barriers against pathogenic microorganisms by killing them, colonising available niches, and consuming and producing nutrients as well as protecting the intestinal mucosa through cellular and/or humoral mechanisms. Evaluation of the dynamics of the host microbiota also suggests that the normal intestinal microbiota interacts with integrative areas of the host's brain via components of the autonomic nervous system (enteric, afferent and efferent pathways) and with target systemic organs via the circulatory and endocrine systems. Various factors can lead to disruption of the gut microbiota (dysbiosis), including acute and chronic infections of the gastrointestinal tract, antibiotic therapy, systemic diseases, ageing, diet and lifestyle (Pimenta et al. 2018).

With age, the body becomes less efficient at managing environmental stressors, leading to a myriad of physiological imbalances, including elevated inflammation, oxidative stress, metabolic dysregulation and mitochondrial damage. There will never be a single therapeutic or dietary solution to manage the increasing chronic diseases associated with ageing; however, maintaining a healthy gut microbiota through the use of probiotic and prebiotic supplements can delay the onset of chronic disease and promote longevity by simultaneously affecting each of the main triggers of ageing. In the present study, an optimised probiotic formulation containing three bioactive probiotics was combined with a new prebiotic triphala rich in polyphenols to create a new symbiotic formulation that promotes longevity. The symbiotic formulation has a combinatorial effect on several ageing markers, including the basic signalling pathways that manage the cross-regulation of these markers. By understanding how gut microbiota and probiotic treatments intersect with these key ageing pathways, specific formulations can be prescribed early in life to prevent the onset of chronic diseases, including cardiovascular disease, diabetes, obesity, cancer and even neurodegeneration (Westfall et al. 2018).

Oxidative stress defines a condition in which the pro-oxidant-antioxidant balance in the cell is disturbed, resulting in DNA hydroxylation, protein denaturation, lipid peroxidation and apoptosis, jeopardising cell viability. Probiotics are known for many beneficial health effects, and the consumption of probiotics alone or in foods shows that strain-specific probiotics can exhibit antioxidant activity and reduce the damage caused by oxidation. Probiotics can modulate the redox status of the host through their metal ion chelating capacity, antioxidant systems and regulatory signalling pathways (Wang et al. 2017).

Several factors are vital for normal intestinal motility including immune and nervous system function, bile acid metabolism and mucus secretion, and gastrointestinal microbiota and fermentation; an imbalance or dysfunction in any of these components can contribute to aberrant intestinal motility and, consequently, constipation symptoms. The luminal intestinal environment, the immune system, the enteric nervous system and the central nervous system are highly interrelated and control intestinal motility; disturbances in any of these overlapping systems can contribute to constipation symptoms. Modifying the intestinal luminal environment with certain probiotic species and strains can affect motility and secretion in the gut and therefore provide benefits for patients with constipation (Dimidi et al. 2017).

The gut microbiome comprises all the microorganisms and their genomes that inhabit the intestinal tract. It is a key node in the bidirectional bi-intestinal axis that develops through early colonisation and through which the brain and gut jointly maintain an organism's health. The bacteria most often exploited as probiotics are the

Gram-positive families of Bifidobacterium and Lactobacillus. Bifidobacteria and Lactobacilli do not have pro-inflammatory lipopolysaccharide chains, and so their propagation in the intestine does not trigger full-blown immunological reactions. With the presence of these bacteria, the immune system learns to distinguish between pro- and anti-inflammatory entities and develops appropriate immunogenic responses. The psychophysiological effects of psychobiotics fall into the following three categories: (i) Psychological effects on emotional and cognitive processes. (ii) Systemic effects on the HPA axis and the glucocorticoid stress response, and inflammation which is often characterised by aberrant cytokine concentrations. Pro-inflammatory cytokines share a strong and well-studied positive association with psychiatric conditions such as depression (Sarkar et al. 2016).

SUMMARY FOR PROBIOTICS

The imbalance of the intestinal microbiota or dysbiosis plays a significant role in the pathogenesis of intestinal disorders.

The studies provide convincing information for the use of probiotics from pregnancy to late adulthood.

Probiotics can be effective in optimising the development of the neonatal microbiome, as the foetus and placenta are not sterile and bacteria are transferred from mother to foetus during pregnancy.

Whether in childhood, adolescence, adults or the elderly, studies show that the prescription of probiotics is necessary in all situations involving microbiota impairment.

In constipation, there may be doubts about the mechanisms of action, but not about the effects for the treatment of constipation. In addition to physiological disorders, probiotics can alleviate the symptoms of depression and anxiety caused by constipation.

Probiotics are defined as live microorganisms that, when administered in adequate quantities, confer a health benefit on the host. Probiotic microorganisms are mainly used as treatments because they can stimulate changes in the intestinal microbial ecosystem and improve the host's immune status.

The documented effects are diverse and the beneficial impact of probiotics is mediated by different mechanisms.

The bacteria most often used as probiotics are the Gram-positive families of Bifidobacterium and Lactobacillus.

COMMENT

To date, there is no evidence that the use of probiotics is unsafe. However, there are also pro-inflammatory and anti-inflammatory foods, high, medium and low glycaemic index foods, nutrient-drug and nutrient-nutrient interactions, etc. For probiotics, it is important that the scientific community continues to endeavour to study the interactions and mechanisms of action of probiotics.

THE INFLUENCE OF DIET ON CONSTIPATION

Diet is believed to be a powerful factor influencing the composition and metabolic activity of an individual's microbiota. The composition of the microbiota in babies changes after weaning, and in adults it varies according to geographical regions due to differences in the food consumed, the type of meat consumed and cooking methods (whether the food is fried or boiled). Therefore, any dietary strategy aimed at modifying the microbiota must be compatible with the individual, as different microbial species respond to different types of dietary components. Although it has been suggested that some patients with irritable bowel syndrome may benefit from dietary fibre, many patients report an increase in abdominal distension and bloating as a result of fibre fermentation (Lee and Lee, 2014).

According to Asakura et al. 2017, Higher carbohydrate intake was marginally associated with a higher prevalence of constipation. Consumption of potatoes, pulses, vegetables and fruit decreased the prevalence of constipation, while higher rice intake was significantly and independently associated with higher prevalence of constipation. Study of preschoolers in Japan.

Many of the commonly used bulking agents are not always effective or well tolerated. There is a need for safe and effective laxation products that will be well accepted and consistently consumed. Whereas the benefits of insoluble fibre for laxation have been equivocal, a meta-analysis indicates that supplementing the diet with soluble fibre decreases the severity of constipation. Various forms of bulk-forming insoluble fibre (e.g. cellulose) can potentially complicate constipation (Buddington et al. 2017).

Dietary fibre has long been used to treat various gastrointestinal conditions. It is widely believed that irritable bowel syndrome is mainly caused by a deficient intake of dietary fibre. Increasing dietary fibre intake has been the standard recommendation for patients with irritable bowel syndrome. However, a systematic meta-analysis based on 12 small studies showed that increasing dietary fibre intake by patients with irritable bowel syndrome did not improve symptoms compared to placebo or a low-fibre diet. Other studies have shown that while consumption of water-insoluble fibre does not improve symptoms of irritable bowel syndrome, consumption of soluble fibre does improve overall symptoms. Dietary fibres can be divided into soluble types (i.e. dissolving in water) and insoluble types based on their physical and chemical properties. Soluble dietary fibre can be subdivided into viscous (gel-forming). Dietary fibre can be divided into short- and long-chain carbohydrates, and fermentable or non-fermentable. Short-chain, soluble and highly fermentable dietary fibre (e.g. oligosaccharides) results in the rapid production of gas that can exceed the capacity of the gastrointestinal tract to absorb gas into the bloodstream for final elimination through the lungs. This imbalance can cause pain, abdominal discomfort, bloating and flatulence. On the other hand, long-chain, intermediate viscous, soluble and moderately fermentable dietary fibre (e.g. psyllium) results in low gas production and the absence of symptoms related to excessive gas production. Insoluble dietary fibre increases faecal mass and speeds up intestinal transit via stimulation, mechanical irritation of the colonic mucosa, with increased secretion and peristalsis. Soluble dietary fibre is fermented by bacteria in the large intestine, which increases stool volume by increasing biomass by fermentation by-products such as gas and short-chain fatty acids. The time and sensation of oro-anal transit are affected by these changes and probably also by other effects on the microbiota, immune cells, intestinal endocrine cells, the enteric nervous system and permeability. Soluble viscous dietary fibre (e.g. psyllium) is minimally fermented and forms a gel that is preserved during its passage through the large intestine and normalises stool form. Probable mechanisms by which dietary fibre affects the functions of the gastrointestinal tract. Dietary fibre acts as a prebiotic for the intestinal microbiota, causing changes in its composition and inducing the growth of beneficial bacteria. The intestinal microbiota, in turn, causes fermentation of the dietary fibre, producing gas, short-chain fatty acids and other by-products. Gas production increases faecal mass and increases luminal pressure. These mechanisms, together with the decrease in luminal pH, stimulate the secretion of serotonin from the EC cell. Serotonin plays an important role in visceral sensitivity. Short-chain fatty acids act on intestinal endocrine cells and/or neurons of the enteric nervous system to alter gastrointestinal motility and secretion. Short-chain fatty acids also act on the cells of the immune system and therefore reduce inflammation. EC-enterochromaffin cell; ISNF, intrinsic sensory nerve fibres; ESNF, extrinsic sensory nerve fibres. In addition, the fermentation of dietary fibre by-products such as short-chain fatty acids (acetate, propionate and butyrate) and the decrease in luminal colonic pH promote the growth of beneficial bacteria such as lactobacilli.

and bifidobacteria. Butyrate is one of the short-chain fatty acids produced by the fermentation of dietary fibre. It has recently been reported that butyrate suppresses colonic inflammation in two ways: i) by inducing T-cell apoptosis, thus eliminating the source of inflammation, and ii) by suppressing interferon-y (IFN-y) mediated inflammation.

The different types of dietary fibre exhibit marked differences in physical and chemical properties, and not all types of fibre are beneficial for patients with irritable bowel syndrome. A general recommendation to increase fibre intake in this group of patients would be inappropriate, as it could worsen symptoms. Long-chain, viscous intermediate, soluble and moderately fermentable dietary fibre (e.g. psyllium) has documented effects in the management of irritable bowel syndrome and can improve patients' overall symptoms. Supplementation with this type of dietary fibre should be recommended for patients with all subtypes of irritable bowel syndrome, namely irritable bowel syndrome with a predominance of diarrhoea, alternating diarrhoea and constipation and irritable bowel syndrome with a predominance of constipation. When starting a fibre supplementation regime, there may be a transient period of abdominal distension, discomfort and altered bowel habits. Fibre supplementation should therefore be started gradually, with consumption increased by no more than 5 g/day per week (El-Salhy et al. 2017).

It is clear that the prescription of insoluble fibres, especially wheat bran, is contraindicated for intestinal syndromes. Wheat bran is used on a large scale by the food industry with the promise that it is wholemeal or partially wholemeal. As well as being untrue, wheat bran is a negative nutrient chelator. In other words, it reduces the absorption of micronutrients. My recommendation is that in order to benefit from fibre, it should come from natural sources such as unripe fruit, cereals such as oats, nuts (chestnuts, almonds, etc.), and vegetables. In addition, fibre sources should be chosen according to each patient's symptoms.

THE INFLUENCE OF PHYSICAL EXERCISE ON CONSTIPATION

Physical activity can have a beneficial effect on a variety of gastrointestinal diseases, however, this effect depends on the mode of exercise, duration and intensity. On the other hand, there is considerable evidence that high-intensity training or prolonged endurance training can exert a negative influence on the gastrointestinal tract, resulting in the exacerbation of symptoms (Bilski et al. 2018)

Constipation is common all over the world, affecting all ages, with prevalence rates of 0.7 to 29.6 per cent in children and adolescents. Physical and emotional well-being is affected and treatment can be expensive. Previously identified risk factors for constipation include female gender, older age, poor socioeconomic status, insufficient fruit and vegetable consumption, psychological problems and insufficient physical activity. Insufficient physical activity and excessive sedentary behaviour were positively associated with constipation in a dose-response relationship. If the association is causal, promoting physical activity and reducing sedentary behaviour could help alleviate constipation in adolescents. Parents and primary care physicians can help adolescents identify constipation and advise on lifestyle changes, including an increase in physical activity (Huang et al. 2014).

The colonic transit time of psychiatric patients was reduced due to increased physical activity through a 12-week combined exercise programme (Song et al. 2018).

It's a shame that the scientific community doesn't study the effects of exercise on constipation. We know that exercise is a powerful remedy for preventing and controlling many pathologies, so why isn't it being studied for constipation?

In Probiotics and Physical Exercise in Constipation, 2009, we said that exercise can improve intestinal motility, but the studies were few and inconclusive. Now, in 2018, the situation continues.

Despite the lack of studies, there is a high degree of confidence on the part of physical educators, nutritionists, acupuncturists and doctors in the effects of exercise on constipation.

In particular, in my practice, I see that the absence or presence of exercise interferes with the aetiology of constipation.

BASED ON PRACTICE AND LONG-STANDING STUDIES

I recommend it:

If blood is a transporting river, it needs water to maintain its bed and consistency. Water is the vehicle that transports heat from the inside of the body to the outside through perspiration. Ingesting too little water retains heat and causes dryness. And according to Chinese medicine, dryness is one of the factors that causes constipation.

A machine can be replaced, while the body needs constant maintenance. Cognitive and motor functions are dependent on neurotransmitters such as dopamine, and a significant tool in maintaining neurotransmitters is physical exercise. Without exercise, the body stiffens and loses mobility. Intestinal motility is indirectly dependent on exercise, which acts on the upper brain and interferes with the lower brain, which is the intestines.

Nutrition is made up of macronutrients and micronutrients whose function is to build, repair, regulate and produce energy. In the hustle and bustle of the modern world, nutrition is more about satisfying hunger than nourishing. Proper nutrition is a mix of foods that come from under and above the ground and from the waters.

Emotional state is a phenomenon linked to the brain, but it is not chemical. It is immaterial and plays an important role in intestinal syndromes. According to Chinese medicine, sadness is in Metal (Lung/ Large Intestine). Fear is in Water (Kidney/Bladder). Anger in Wood (Liver/Gallbladder). Joy in Fire (Heart/Small Intestine/Heating Gland/Pericardium or Sex Circulation). Worry in Earth (Spleen/Pancreas). The Five Elements work together to nourish and control. It's like a family, if one has a problem everyone is affected. If the Earth Element is affected by worry, it goes into deficiency, which in turn does not nourish Metal. Unnourished Metal doesn't control Wood. Wood in excess feeds Fire too much. High Fire produces a lot of heat and dries out Water. Heat and little water dries out the intestines. A weakened Spleen, which is responsible for most of the digestive dynamics, will result in constipation and haemorrhoids.

It is clear from Chinese medicine that the end product of the aetiology, in simplified form, is a mechanical failure due to energy stagnation resulting from the deficiency or excess of some Element. In particular, Wood and Earth.

CHAPTER 5

GENERAL SUMMARY

Constipation is a gastrointestinal disorder with a high prevalence worldwide. It affects everyone from newborns to the elderly. The aetiology is multifactorial and causes physical and psychological suffering.

In oriental medicine, pathologies are considered to be energy imbalances in a particular element within the physiology of the 5 Elements, and each Element corresponds to an organ or viscera. In the case of constipation, it is a deficiency in the Earth Element (Spleen), which is responsible for transformation and transport.

One of the notable characteristics of constipation is that it triggers other pathologies, as many pathologies trigger constipation. Another relevant situation reported in the studies is the indiscriminate or prescribed use of medication that triggers constipation and other illnesses.

The conventional therapeutic approach to treating constipation is lifestyle change and laxatives as the basis of pharmacological treatment. In most cases, pharmacological treatment causes more discomfort than benefit.

ACUPUNCTURE AND PROBIOTICS AS A PROPOSAL FOR THE TREATMENT OF CONSTIPATION

Both acupuncture and probiotics have a long and effective history of treating various syndromes, including constipation.

Acupuncture is the gold standard of Chinese medicine and has been revised and refined by Japanese and Korean medicine. Acupuncture treatment involves the insertion of fine needles into the skin and underlying muscle and the needles are stimulated manually or electrically. Its mode of action at a biochemical level, acupuncture stimulates neurotransmitters that are important markers for intestinal health and improved mood.

Lactobacillus and bifidobacterium increased with acupuncture, indicating a reduction in dysbiosis.

Acupuncture restores coordination between the contraction and relaxation muscles, improving intestinal motility.

Of all the complementary medical systems, acupuncture enjoys the greatest credibility. Its healing effect has been widely recognised and accepted worldwide by the international medical community.

Probiotics are live microorganisms which, when administered in adequate quantities, confer health benefits on the host.

The intestines are colonised by trillions of microorganisms and play an important role in health and disease. The composition and activity of the intestinal microbiota can change dramatically due to factors such as stress, diet and drugs. Certain drugs decrease beneficial bacteria and increase pathogenic ones. Through the gut-brain axis, the intestinal microbiota regulates states of anxiety, cognition, mood and pain.

The main benefit of probiotics is that they contribute to a balanced microbiota. By creating a favourable intestinal environment, probiotics promote intestinal and immune system health. Due to their safety and healing effect, probiotics have been widely accepted. In recent years, probiotic microorganisms are being used as a treatment because they can stimulate changes in the intestinal microbial ecosystem and improve the patient's immune status.

The beneficial impact of probiotics is mediated by different mechanisms. Probiotics increase the function of the intestinal barrier, prevent bacterial displacement and modulate inflammation via immune receptors and regulate the expression of intestinal genes. The intestinal microbiota interacts with different organs and, while the natural community of microorganisms contributes to health, intestinal dysbiosis is linked to the onset or worsening of various diseases.

Various factors can lead to dysbiosis, such as acute or chronic infections of the gastrointestinal tract, antibiotic therapy, systemic diseases, ageing, diet and lifestyle.

The intestinal environment, the immune system, the enteric nervous system and the central nervous system are highly interrelated and control intestinal motility;

disturbances in any of these overlapping systems can contribute to constipation. Modifying the intestinal environment with probiotics can positively affect motility and secretion in the gut and provide benefits for patients with constipation.

The bacteria most often exploited as probiotics are the Gram-positive families of Bifidobacterium and Lactobacillus. With the presence of these bacteria, the immune system learns to distinguish between pro- and anti-inflammatory entities and develops appropriate immunogenic responses

So, based on scientific evidence and practical experience as a nutritionist and acupuncturist, I can say that acupuncture and probiotics are effective in treating the suffering caused by constipation symptoms.

Comments:

1- There are cases where constipation is caused by an anatomical defect. At the doctor's discretion, surgery may be indicated.

2- Constipation caused by medication: Acupuncture and probiotics for the duration of the drug intervention.

3- Constipation caused by physical inactivity (bedridden patients/neurological pathologies): Constant probiotics and acupuncture whenever necessary to release stagnant faeces.

PRESCRIPTION

Qualitative: at the moment the most suitable are from the Lactobacillus and Bifidobacterium families. Examples: Lactobacillus acidophillus, Lactobacillus casei, Lactobacillus rhamnosus, Lactobacillus reuteri, Lactobacillus plantarum, Bifidobacterium lactis and Bifidobacterium brevis.

Quantitative: Can be according to the severity of symptoms. Dose response. For example, severe discomfort, start with five billion/day. At least the types described above.

In children, if in doubt, ask the pharmacist for help.

For acupuncture, use the points that appear most frequently in studies. These are IG11 and E37. Plus points according to energy assessment. BP15 can also complement the treatment.

About the author

Antonio Francisco Moreira

Nutritionist and Acupuncturist

Graduated in Nutrition, Postgraduate in Obesity and Slimming

Specialist in Acupuncture and Electroacupuncture

A student of physical exercise for over 30 years.

Quantum therapy student

Co-author of the book AAA Diet and its approach in oriental dietetics

Email: antonionut12@hotmail.com

September 2018

REFERENCES

Agamennone, V Krul, C.A.M-; Rijkers, G-; Kort, R- **A practical guide for probiotics applied to the case of antibiotic-associated diarrhea in The Netherlands.** BMC Gastroenterol. 2018; 18(1):103. [PubMed]

Asakura, K- Masayasu, S- Sasaki, S. **Dietary intake, physical activity, and time management are associated with constipation in preschool children in Japan.** Asia Pac J Clin Nutr 2017; 26(1):118-129. [PubMed]

Adams-Carr, K.L; Bestwick, J.P-; Shribman, S.; Lees, A Schrag, A Noyce, A.J. **Constipation preceding Parkinson's disease: a systematic review and meta-analysis.** J Neurol Neurosurg Psychiatry. 2016; 87(7):710-6. [PubMed]

Agarwal, R.K. **Probiotics - the health friendly gut bacteria.** Indian Paediatric. New Delhi. Vol. 45. Num.12. 2008. p. 953-954.[PubMed]

Auteroche, B.; Navailh, P. **Diagnosis in Chinese Medicine**. Andrei. 1992; São Paulo.

Black, C.J.; Ford, A.C. **Chronic idiopathic constipation in adults: epidemiology, pathophysiology, diagnosis and clinical management.** Med J Aust. 2018; 209(2):86-91. [PubMed]

Bilski, J.; Mazur-Bialy, A Magierowski, M.; Kwiecien, S.; Wojcik, D.; Ptak- Belowska, A Surmiak, M.; Targosz, A Magierowska, K Brzozowski, T. **Exploiting significance of physical exercise in prevention of gastrointestinal disorders.** Curr Pharm Des 2018. doi: 10.2174/1381612824666180522103759. [PubMed]
Bomfim, I.Q.M.; Nunes, L.S.; Alves, T.C. **Prevalence of constipation in physiotherapy students at a university in Maceió/AL.** Revista Ciênc. Méd.Biol. v. 16. n.1. p. 79-84, 2017.

Bu, F.Y.; Wei, F.; Dai, X.S.; Tan, Q.W.; Cheng, B. **Progress on the Relationship Between the Regulation of Gastrointestinal System Function by Acupuncture and Central Nuclei.** Zhen Ci Yan Jiu. 2017; 25, 42 (4): 363-6 [PubMed]

Buddington, R.K; Kapadia, C; Neumer, F; Theis, S **Oligofructose Provides Laxation for Irregularity Associated with Low Fiber Intake.** Nutrients. 2017; 9(12):E1372. [PubMed]

Baçturk, A; Artan, R.; Yilmaz, A. **Efficacy of synbiotic, probiotic, and prebiotic treatments for irritable bowel syndrome in children: A randomised controlled trial.** Turk J Gastroenterol. 201627(5):439-443 [PubMed].

Broide, E.; Pintov, S.; Portnov, S.; Barg, J.; Klinowski, E.; Scapa, E. **Effectiveness of acupuncture for treatment of childhood constipation.** Dig Dis Sci. 2001; 46(6): 1270-6 [PubMed]

Bekkali, N-L-H.; Bongers, M.E.J.; Van den Berg, M.M.; Liem, O.; Benninga, M.A. **The role of a probiotics mixture in the treatment of childhood constipation: a pilot study.** Nutrition Journal. London. Vol. 6. Num.17. 2007. p. 1-6. [PubMed]

Carter, D.; Bardan, E.; Dickman, R. **Comparison of strategies and goals for treatment of chronic constipation among gastroenterologists and general practitioners.** Ann Gastroenterol. 2018; 31(1):71-76 [PubMed]

Christie, J.; Shroff, S.; Shahnavaz, N.; Carter, L.A.; Harrison, M.S.; Dietz-Lindo, K.A.; Hanfelt, J.; Srinivasan, S. **A Randomised, Double-Blind, Placebo- Controlled Trial to Examine the Effectiveness of Lubiprostone on Constipation Symptoms and Colon Transit Time in Diabetic Patients.** Am J Gastroenterol. 2017; 112(2):356-364.[PubMed]

Chen,Q.; Jiang, J.**Relationship between functional constipation and brain gut-microbiota axis.**Zhonghua Wei Chang Wai Ke za Zhi. 2017; 25;20(12): 1345-1347 [PubMed]

Chen, L.; Jin, X.; Jiang, X.; Wang, C.; Shi, Y; Wang, L. **Acupoint embedding for female functional constipation:a randomised controlled trial.** Zhonqquo Zhen Jiu. 2017; 12;37(7): 717-721 [PubMed]

Cao, H.; Liu, X.; An, Y.; Zhou, G.; Liu, Y.; Xu, M.; Dong, W.; Wang, S.; Yan, F.; Jiang, K.; Wang, B. **Dysbiosis contributes to chronic constipation development via regulation of serotonin transporter in the intestine.** Sci Rep. 2017; 7(1): 10322 [PubMed]

Chey, W.D.; Camilleri, M.; Chang, L.; Rikner, L.; Graffner, H. **A randomised placebo-controlled phase IIb trial of a3309, a bile acid transporter inhibitor, for chronic idiopathic constipation.** Am J Gastroenterol. 2011; 106(10): 1803-1812 [PubMed]

Cabyoglu, M.T.; Ergene, N.; Tan, U. **The mechanism of acupuncture and clinical applications.** Int J Neurosci. 2006; 116(2): 115-125 [PubMed]

Diaz, S.; Mendez, M.D. Constipation. StatPearls [Internet]. Treasure Island (FL): StatPearls Publishing; 2018 [PubMed]

Dimidi, E.; Christodoulides, S.; Scott, S.M- Whelan, K- **Mechanisms of Action of Probiotics and the Gastrointestinal Microbiota on Gut Motility and Constipation.** Adv Nutr. 2017; 8(3):484-494 [PubMed].
Del'arco, A.P.W.T Magalhães, P; Quilici, F.A. **SIM BRASIL STUDY - WOMEN'S GASTROINTESTINAL HEALTH: GASTROINTESTINAL SYMPTOMS AND**

IMPACT ON THE BRAZILIAN WOMEN QUALITY OF LIFE. Arg Gastroenterol. 2017; 54(2):115-122. [PubMed]

Du, W.F.; Yu, L.; Yan, X.K.; Wang, F.C. **Met-analysis on randomised controlled clinical trials of acupuncture and moxibustion on constipation.** Zhonqquo Zhen Jiu. 2012; 32(1): 92-96 [PubMed]

Drouault-Holowacz, S.; Bieuvelet, S.; Burckel, A.; Cazaubiel, M.; Dray, X.; Marteau, P. **A double blind randomised controlled trial of a probiotic combination in 100 patients with irritable bowel syndrome.** Gastroentérologie Clinique et Biologique. Paris. Vol. 32. Num. 2. 2008. p.147- 152. [PubMed]

Dughera, L.; Elia, C.; Navino, M.; Cisarò, F. **Effects of symbiotics preparations on constipated irritable bowel syndrome symptoms.** Actabiomedica. Parma. Vol 78. Num. 2. 2007. p. 111-116. [PubMed]

El-Salhy, M.- Ystad, S.O- Mazzawi, T Gundersen, D- **Dietary fibre in irritable bowel syndrome (Review).** Int J Mol Med. 2017; 40(3):607-613 [PubMed].

Forootan, M.; Bagheri, N.; Darvishi, M. **Chronic constipation: A review of literature.** Medicine (Baltimore). 2018;97(20): e 10631[PubMed]

Fuller, R. **Probiotics in man and animals.** Journal of Applied Bacteriology. Oxford. Vol. 66. Num. 5. 1989. p. 365-378. [PubMed]

Felippi, R.S.R.; Franca, Rita; Silva, L.R.; Marques, C.D.F. Prevalence of constipation in obese patients followed up at the outpatient clinic of a paediatric hospital. Journal of Medical and Biological Sciences. 2014; 13(2): 152-155
Gao, J-; Xu, K Liu, H Liu, G Bai, M-; Peng, C Li, T Yin, Y **Impact of the Gut Microbiota on Intestinal Immunity Mediated by Tryptophan Metabolism.** Front Cell Infect Microbiol. 2018; 8:13. [PubMed]

Gan, J-; Wan, Y- Shi, J-; Zhou, M-; Lou, Z-; Liu, Z- **A survey of subjective constipation** in Parkinson's disease patients in shanghai and literature review. BMC Neurol. 2018; 18(1):29. [PubMed]

Grossi, U.; Stevens, N.; McAlees, E.; Lacy-Colson, J.; Brown, S.; Dixon, A.; Di Tanna, G.L.; Scott, S.M.; Norton, C.; Marlin, N.; Mason, J.; Knowles, C.H. **Stepped-wedge randomised trial of laparoscopic ventral mesh rectopexy in adults with chronicconstipation: study protocol for a randomized controlled trial.** Trials. 2018; 19(1): 90 [PubMed]

Gao, Y.; Li, J.; Su, M.; Li, Y. **Acupuncture with smoothing liver and regulating qi for post-stroke slow-transit constipation and its gastrointestinal hormone level.** Zhonqquo Zhen Jiu. 2017; 37(2):125-129. [PubMed]

Gupta, V Nag, D- Garg, P- **Recurrent urinary tract infections in women: How promising is the use of probiotics?** Indian J Med Microbiol. 2017; 35(3):347-354. [PubMed]

Ge, X.; Ding, C.; Gong, J.; Tian, H.; Wei, Y.; Chen, Q.; Gu, L.; Li, N. **[Short-term efficacy on faecal microbiota transplantation combined with soluble dietary fibre and probiotics in the treatment of slow transit constipation].** Zhonghua Wei Chang Wai Ke Za Zhi. 2016; 19(12):1355-1359. [Pubmed]

Gasbarrini, G- Bonvicini, F.; Gramenzi, A. **Probiotics History.** J Clin Gastroenterol. 2016; 50(2):116-119. [PubMed]
Gallegos-Orozco, J.F.; Foxx-Orenstein, A.E.; Sterler, S.M.; stoa, J.M. **Chronic constipation in the elderly.** Am J Gastroenterol. 2012; 107(1): 1825 [PubMed]

Gibson, G.R.; Fuller, R. **Aspects of in vitro and in vivo research approaches directed toward indentifying probiotics and prebiotics for human use.** The Journal of Nutrition. Bethesda. Vol. 130. Num. 2. 2000. p. 391-395. [PubMed]

Huang, L- Zhu, Q- Qu, X- Qin, H- **Microbial treatment in chronic constipation.** Sci China Life Sci. 2018; 61(7):744-752. [PubMed]

Handaya, Y.; Maryanto, A.; Marijata. **Side-to-Side Ileosigmoidostomy Shunting Surgery for the Treatment of Elderly Patients With Chronic Constipation.** Ann Coloproctol. 2017; 33(6): 249-252 [PubMed]

Huang, R.; Zhao, J.; Wu, L.; Dou, C.; Liu, H.; Weng, Z.; Lu, Y.; Shi , Y.; Wang, X.; Zhou, C.; Wu, H. **Mechanisms underlying the analgesic effect of moxibustion on visceral pain in irritable bowel syndrome: a review.** Evid Based Complement Alternat Med. 2014; 2014: 895914 [PubMed]

Huang, R.; Ho, S.Y.; Lo, *W.S.;* Lam, T.H- **Physical activity and constipation in Hong Kong adolescents. Plos One.** 2014; 9(2): e90193. [PubMed]

He, W.; Tong, Y.Y.; Zhao, Y.K.; Rong, P.J.; Wang, H.C. **Bibliometrics study on indications of acupuncture therapy based on foreign acupuncture clinicaltrials.** Zhen Ci Yan Jiu. 2012; 37(5): 428-430 [PubMed]

Hicks, A.; Hicks, J.; Mole, P. **Five Elements Constitutional Acupuncture.** Roca. São Paulo, 2007
Jesus, F.R.; Diniz, J.C. **Prevalence of constipation in the elderly: an association with its triggering factors.** Brazilian Journal of Life Sciences. v. 5, n2, 2017.

Kassaian, N Feizi, A; Aminorroaya, A.; Amini, M. **Probiotic and synbiotic supplementation could improve metabolic syndrome in prediabetic adults: A randomised controlled trial.** Diabetes Metab Syndr. 2018. pii: S1871-4021(18)30283-2. [PubMed]

Kirihara, N.; Kamitomo, M.; Tabira, T.; Hashimoto, T.; Taniguchi, H.; Maeda, T. **Effect of probiotics on perinatal outcome in patients at high risk of preterm birth.** J Obstet Gynaecol Res. 2018; 44(2):241-247. [PubMed]

Kearney, R.; Edwards, T Bradford, M.; Klein. E- **Emergency Provider Use of Plain Radiographs in the Evaluation of Paediatric Constipation.** Paediatr Emerg Care. 2018 doi: 10.1097/PEC.0000000000001549. [PubMed]

Koppen, I.J.N.; Saps, M.; Lavigne, J.V.; Nurko, S.; Taminiau, J.A.J.M.; Di, L.C.; Benninga, MA **Recommendations for pharmacological clinical trials in children with functional constipation: The Rome foundation paediatric subcommittee on clinical trials.** Neurogastroenterol Motil. 2018. doi: 10. 1111/nmo. 13294 [PubMed]

Kanmani, P·; Kim, H- **Protective Effects of Lactic Acid Bacteria Against TLR4 Induced Inflammatory Response in Hepatoma HepG2 Cells Through Modulation of Toll-Like Receptor Negative Regulators of Mitogen-Activated Protein Kinase and** NF-κB **Signaling.** Front Immunol. 2018; 9:1537. [PubMed]

Karkow, F.J.A.; Faintuch, J.; Karkow, A.G.M. **Probiotics: medical perspectives.** Journal of the Rio Grande do Sul Medical Association. Porto Alegre. Vol. 51. Num. 1. 2007. p. 38-48. [PubMed]
Liang, C Wang, K.Y Gong, M.R Li, Q· Yu, Z. Xu, B **Electro- acupuncture at ST37 and ST25 induce different effects on colonic motility via the enteric nervous system by affecting excitatory and inhibitory neurons.** Neurogastroenterol Motil. 2018; 30(7):e13318. [PubMed]

Liu, X.Y· Dai, X.H· Zou, W· Yu, X.P· Teng, W· Wang, Y· Yu, W.W· Ma, H.H· Chen, Q.X· Liu, P· Guan, R.Q· Dong, S.S. **Acupuncture through** *Baihui* **(DU20) to** *Qubin* **(GB7) mitigates neurological impairment after intracerebral haemorrhage.** Neural Regen Res. 2018; 13(8):1425-1432. [PubMed]

Leblhuber, F· Steiner, K· Schuetz, B· Fuchs, D· Gostner, J.M **Probiotic Supplementation in Patients with Alzheimer's Dementia - An Explorative Intervention Study.** Sed toCurr Alzaheimer Res. 2018. doi: 10.2174/1389200219666180813144834. [PubMed]

Lee, H.Y· Kwon, O.J Kim, J.E· Kim, M· Kim, A.R· Park, H.J· Cho, J.H· Kim, J.H· Choi, S.M **Efficacy and safety of acupuncture for functional constipation: a randomised, sham-controlled pilot trial.** BMC Complement Altem Med. 2018; 18(1):186. [PubMed]

Larkin, P.J· Cherny, N.ı-· La Carpia, D· Guglielmo, M· Ostgathe, C· Scotté, F· Ripamonti, C.I **Diagnosis, assessment and management of constipation in** advanced cancer: **ESMO Clinical Practice Guidelines**. Ann oncol. 2018; doi: 10.1093/annonc/mdy148 [PubMed].

Lv, J.Q.; Wang, C.W.; Liu, M.Y.; Zhao, Y.; Wen, Q.; Li, N. **Observation on the Efficacy of Electroacupuncture for Functional Constipation.** Zhen Ci Yan Jiu. 2017; 42(3): 254-258 [PubMed]

Lenoir-Wijnkoop, I Gerlier, L; Roy, D; Reid, G **The Clinical and Economic Impact of Probiotics Consumption on Respiratory Tract Infections: Projections for Canada.** PloS One. 2016; 11(11):e0166232. [PubMed]

Lee, K.N; Lee, O.Y Intestinal microbiota in pathophysiology and management of irritable bowel syndrome. World J Gastroenterol. 2014; 20(27):8886-97. [PubMed]

Li, M.K.; Lee, T.F.; Suen, K.P. **Complementary effects of auricular acupressure in relieving constipation symptoms and promoting disease-specific health-related quality of life: A randomised placebo- controlled trial.**Complement Ther Med. 2014; 22(2): 266-277 [PubMed].

Lin, J.G.; Chou, P.C.; Chu, H.Y. **An exploration of the needling depth in acupuncture: the safe needling depth and the needling depth of clinical efficacy.** Evid Based Complement Alternat Med. 2013; 2013: 740508 [PubMed]

Mohajeri, M.H Brummer, R.J.M- Rastall, RA; Weersma, R.K; Harmsen, H.J.M; Faas, M.; Eggersdorfer, M **The role of the microbiome for human health: from basic science to clinical applications.** Eur J Nutr. 2018. doi: 10.1007/s00394-018-1703-4. [PubMed]

Meng, C; Bai, CBrown, T.D; Hood, L.E.; Tian, Q **Human Gut Microbiota and Gastrointestinal Cancer. Genomicis Proteomics Bioinformatics.** 2018; 16(1):33-49. [PubMed]

Mafi, A; Namazi, G; Soleimani, A; Bahmani, F; Aghadavod, E; Asemi, Z. **Metabolic** and genetic response to probiotics supplementation in patients with diabetic nephropathy: a randomised, double-blind, placebo-controlled trial. Foood Funct. 2018. doi: 10.1039/c8fo00888d. [PubMed]

Mearin, F.; Ciriza, C.; Minquez, M.; Rey, E.; Mascort, J.J.; Pena, E.; Canones, P.; Júdez, J. **Irritable bowel syndrome with constipation and functional constipation in adults: Treatment (Part 2 of 2).** Aten Primaria. 2017; 49(3): 177-194 [PubMed] Miller, L.E.; Ibarra, A.; Ouwehand, A.C.; Zimmermann, A.K. **Normative values for stool frequency and form using Rome III diagnostic criteria for functionalconstipation in adults: systematic review with meta-analysis.** Ann Gastroenterol. 2017; 30(2): 161-167 [PubMed]

Martínez-Martínez, M.I.; Calabuig-Tolsá, R.; Cauli, O. **The effect of probiotics as a treatment for constipation in elderly people: A systematic review.** Arch Gerontol Geriatr. 2017; 71:142-149. [PubMed]

Mezzasalma, V Manfrini, E.; Ferri, E.; Sandionigi, A La Ferla, B.; Schiano, I.; Michelotti, A.; Nobile, V Labra, M.; Di Gennaro, P- **A Randomised, Double-Blind, Placebo-Controlled Trial: The Efficacy of Multispecies ProbioticSupplementation in Alleviating Symptoms of Irritable Bowel Syndrome Associated with Constipation.** Biomed Res Int. 2016; 2016(10):4740907 [PubMed].

McFarland, LV **From yaks to yogurt: the history, development, and current use of probiotics.** Clin Infect Dis. 2015; 60(2):85-90. [PubMed]

Moreira, A.F.; Maciel, M.M.; Navarro, F.; Silva, B.M. **Probiotics and physical exercise in constipation**. Revista Brasileira de Obesidade, Nutrição e Emagrecimento. v. 3, n 16, p.305-315, 2009

McCrea, G.L.; Miaskowski, C.; Stotts, N.A.; Macera, L.; Varma, M.G. **Pathophysiology of constipation in the older adult.** World Journal of Gastroenterology. Dongsihuan Zhonglu. Vol. 14. Num. 17. 2008. p. 26312638. [PubMed]

Ojetti, V.; Petruzziello, C.; Migneco, A.; Gnarra, M.; Gasbarrini, A.; Franceschi, F.

Effect of Lactobacillus reuteri (DSM 17938) on methane production in patients affected by functional constipation: a retrospective study. Eur Rev Med Pharmacol Sci. 2017; 21(7):1702-1708. [PubMed]

Otles, S.; Çagindi, O.; Akçiçek, E. **Probiotics and Health.** Asian Pacific Journal of Cancer Prevention. Bangkok. Vol. 4. Num. 4. 2003. p. 369-372. [PubMed]

Plaza-Díaz, J.; Ruiz-Ojeda, F.J.; Gil-Campos, M.; Gil, A. **Immune-Mediated Mechanisms of Action of Probiotics and Synbiotics in Treating Paediatric Intestinal Diseases.** Nutrients. 2018. 10(1):e42. [PudMed]

Pimenta, F.S.; Luaces-Regueira, M.; Ton, A.M.; Campagnaro, B.P.; Campos-Toimil,M.; Pereira,T.M.; Vasquez, E. C. **Mechanisms of Action of Kefir in Chronic Cardiovascular and Metabolic Diseases.** Cell Physiol Biochem. 2018 48(5):1901-1914. [PubMed]

Pei, L.; Zhu, L.; Sun, J.; Wu, X.; Chen, L. **Constipation predominant irritable bowel syndrome treated with acupuncture for regulating the mind and strengthening the spleen: a randomised controlled trial.** Zhonqquo Zhen Jui. 2015; 35(11): 1095-8 [PubMed]

Qin, Q.G.; Gao, X.Y.; Liu, K.; Yu, X.C.; Li, L.; Wang, H.P.; Zhu, B. **Acupuncture at heterotopic acupoints enhances jejunal motility in constipated and diarrheic rats.** World Gastroenterol. 2014; 20(48): 1827183 [PubMed]

Robin, S.G.; Keller, C.; Zwiener, R.; Hyman, P.E.; Nurko, S.; Saps, M.; Di Lorenzo, C.; Shulman, R.J.; Hyams, J.S.; Palsscn, O. **Prevalence of Paediatric Functional Gastrointestinal Disorders Utilizing the Rome IV Criteria.** J Pediatr. 2018; 3476(17): 31634-7 [PubMed]

Riezzo, G.; Orlando, A.; D'Attoma, B.; Linsalata, M.; Martulli, M.; Russo, F. **Randomised double blind placebo controlled trial on Lactobacillus reuteri DSM 17938: improvement in symptoms and bowel habit in functional constipation.** Benef Microbes. 2018; 9(1): 51-60 [PubMed]

Roo, S.S.; Rattanakovit, K.; Patcharatrakul, T. **Diagnosis and management of chronic constipation in adults.** Nat Rev Gastroenterol Hepatol. 2016; 13(5): 295-305. [PubMed]

Ruiz-Lópes, M.C.; Coss-Adame, E. **Quality of life in patients with different constipation subtypes based on the Rome III criteria**. Rev Gastroenterol Mex. 2015, 80 (1): 13-20 [PubMed]

Shin, J.; Park, H. **Effects of Auricular Acupressure on Constipation in Patients With Breast Cancer Receiving Chemotherapy: A Randomised Control Trial.** West J Nurs Res. 2018; 40(1): 67-83 [PubMed]

Song, B. κ- Kim, Y.S Kim, H.S- Oh, J.W- Lee, q- Kim, J.S. **Combined exercise improves gastrointestinal motility in psychiatric in patients.** World J Clin Cases. 2018; 6(8):2017-213. [PubMed]

Su, A.F.K. **Acupuncture for relief of constipation**. Kimg Acupuncture and Traditional Chinese Medicine. 2017

Sharma, A.; Rao, S. **Constipation: Pathophysiology and Current Therapeutic Approaches.** Handb Exp Pharmacol. 2017; 239-59-74 [PubMed]

Scourboutakos, M.J-; Franco-Arellano, B- Murphy, S.A Norsen, S- Comelli, E.M- L'Abbé, M.R- **Mismatch between Probiotic Benefits in Trials versus Food Products.** Nutrients. 2017; 9(4):E400 [PubMed].

Sarkar, A Lehto, S.M- Harty, s- Dinan, T.G-; Cryan, J.F- Burnet, P.W.J- **Psychobiotics and the Manipulation of Bacteria-Gut-Brain Signals.** Trends Neurosci. 2016; 39(11):763-781.[PubMed]
SanfAnna, M.S.L.; Ferreira, C.L.L.F. Prevalence of constipation in the municipality of Viçosa/Minas Gerais. Nutrição Brasil.v.15, n.1, 2016.

Schmidt, F.M.Q.; Santos, V.L.C.G.; Domansky, R.C.; Barros, E.; Bandeira, M.A.; Tenório, M.A.M.; Jorge, J.M.N. **Prevalence of self-reported constipation in adults from the general population**. Revista da Escola de Enfermagem da USP. 2015; 49(3): 443-452

Schrezenmeir, J.; Vrese, M. **Probiotics, prebiotics, and symbiotics- approaching a definition.** The American Journal of Clinical Nutrition. Houston. Vol.73. Num. supp.1. 2001. p. 361-364. [PubMed]

Tantawy, S.A.; Kamel, D.M.; Abdelbasset, *W.K*; Elgohary, H.M- **Effects of a proposed physical activity and diet control to manage constipation in middle-aged obese women.** Diabetes Metb Syndr. 2017; 10:513-519 [PubMed].

Tambucci, R.; Quitadamo. PThapar, N.; Zenzeri, L; Caldaro. T Staiano. A Verrotti. A Borrelli, O- **Diagnostic tests in paediatric constipation.** J Pediatr Gastroenterol Nutr. 2017; .doi: 10.1097/MPG.0000000000001874. [PubMed]

Tjen-A-Looi, S.C.; Fu, L.W. **Sustained effects of acupuncture in treatment of chronic constipation.** Ann Palliat Med. 2017; 6(2): 124-127 [PubMed]

Taba Taba Vakili, S; Nezami, B.G- Shetty, A Chetty, V.K.; Srinivasan, S- **Association of high dietary saturated fat intake and uncontrolled diabetes with constipation:** evidence from the National Health and Nutrition Examination Survey. **Neurogastroenterol Motil. 2015; 27(10):1389-97. [PubMed]**

Turan, |- Dedeli, O.; Bor, S.; Ilter, T. **Effects of a kefir supplement on symptoms, colonic transit, and bowel satisfaction score in patients with chronic constipation: a pilot study.** Turk J Gastroenterol. 2014; 25(6):650- 6. [PubMed]
Takahashi, T. **Effect and mechanism of acupuncture on gastrointestinal diseases.** Int Rev Neurobiol. 2013; 111-273-294 [PubMed]

Uchiyama-Tanaka, Y- **Case Study of Homeopathic Bowel Nosode Remedies for Dysbiotic Japanese Patients.** J Altem Complement Med. 2018; 24(2):187-193. [Pubmed]

Vespasiani-Gentilucci, U; Gallo, P; Picardi, A. **The role of intestinal microbiota in the pathogenesis of NAFLD: starting points for intervention.** Arch Med Sci. 2018; 14(3):701-706. [PubMed]

Veiga, D.R- Mendonça, L· Sampaio, R·· Lopes, J.C Azevedo, L.F- **Incidence and Health Related Quality of Life of Opioid-Induced Constipation in Chronic Noncancer Pain Patients: A Prospective Multicentre Cohort Study.** Pain Res Treat. 2018; 2018:5704627 [PubMed].

van Summeren, J.J.G.T; Holtman, G.A Lisman-van Leeuwen, Y·· Louer, L.E.A.M-; van Ulsen-Rust, A.H.C Vermeulen, K.M- Kollen, B.J- Dekker, J.H·· Berger, M.Y. **Physiotherapy plus conventional treatment versus conventional treatment only in the treatment of functional constipation in children: design of a randomised controlled trial and cost-effectiveness study in primary care.** BMC Psdiatr. 2018; 18(1):249. [PubMed]

Vitton, V.; Damon, H.; Benezech, A.; Bouchard, D.; Brardjanian, s.; Brochard, C.; Coffin, B.; Fathallah, H.; Hiquero, T.; Jouet, P.; Leroi, A.M.; Luciano, L.; Meurette, G.; Piche, T.; Robert, A.; Sabate, J.M.; Siproudhis, L. **Clinical practice guidelines from the French National Society of Coloproctology in treating chronic constipation.** Eur J Gastroenterol Hepatol. 2018; doi: 10.1097/MEG.0000000000001080. [PubMed]

Valdovinos, M.A Montijo, E·· Abreu, A.T Heller, S González-Garay, A·· Bacarreza, D·· Bielsa-Fernández, M·· Bojórquez-Ramos, M.C Bosques- Padilla, F·· Burguete-García, A.I-; Carmona-Sánchez, R·· Consuelo-

Sánchez, A· ; Coss-Adame, E· ; Chávez-Barrera, JA; de Arino, M· ; Flores- Calderón, J· ; Gómez-Escudero, O· ; González-Huezo, M.S· ; Icaza-Chávez, M.E· ; Larrosa-Haro, A· ; Morales-Arámbula, M· ; Murata, C· ; Ramírez-Mayans, J.A Remes-Troche, J.M· ; Rizo-Robles, T Peláez-Luna, M· Toro-Monjaraz, E.M· ; Torre, A· ; Urquidi-Rivera, M.E· ; Vázquez, R· ; Yamamoto-Furusho, J.K Guarner, F· **The Mexican consensus on probiotics in gastroenterology.**
Rev Gastroenterol Mex. 2017; 82(2):156-178. [PubMed]

Vanderploeg, K.; Yi, X. **Acupuncture in modern society.** J Acupunct Meridian Stud. 2009; 2(1): 26-33 [PubMed]

Wang, L· ; Guo, M.J· ; Gao, Q· ; Yang, J.F· ; Yang, L· ; Pang, X.; L· ; Jiang, X.J- **The effects of probiotics on total cholesterol: A meta-analysis of randomized** controlled trials. Medicine (Baltimore). 2018; 97(5):e9679. [PubMed]

Wen, J. Zhuang,Z· ; Zhao ,M.; Xie, D· ; Xie, B· ; Zhuang, L·; Liang, Z· ;Wu,
W Xu.H- **Treatment of poststroke constipation with moxibustion: A case report.** Medicine(Baltimore). 2018; 97(24):e11134. [Pubmed]

Westfall, S· ; Lomis, N· ; Prakash, S· **Longevity extension in Drosophila through gut-brain communication.** Sci Rep. 2018; 8(1):8362 [PubMed].

Waterham, M.; Kaufman, J.; Gibb, S. **Childhood constipation**. Aust Fam Physician. 2017; 46(12): 908-912 [PubMed]

Wang, B.; Chen, Y.; Chen, S. **Clinical observation of slow transit constipaton treated with acupuncture and modified weitong xiaop formula.** Zhonqquo Zhen Jiu. 2017; 37(2): 130-134 [PubMed]

Wang, Y; Wu, Y; Wang, Y; Xu, H Mei, X; Yu, D.; Wang, Y; Li, W **Antioxidant Properties of Probiotic Bacteria.** Nutrients. 2017; 9(5):E521. [PubMed]
Wu, J.N.; Zhang, B.Y.; Zhu, W.Z.; Du, R.S.; Liu, Z.S. **Comparison of efficacy on functional constipation treated with electroacupuncture of different acupoint prescriptions: a randomized controlled pilot trial.** Zhonqquo Zhen Jiu. 2014; 34(6): 521-8 [PubMed]

White, A. **Western medical acupuncture: a definition.** Acupunct Med. 2009; 27(1): 33-5 [PubMed]

Xu, N· Fan, W.; Zhou, X· Liu, Y Ma, P· Qi, S· Gu, B· **Probiotics decrease depressive behaviours induced by constipation via activating the AKT signaling pathway.** Metb Brain Dis. 2018. doi: 10.1007/s11011 -018-0269-4. [PubMed]

Xu, J.; Du, Y.; Zhang, X.; Zhang, X.; Xing, H.; Pan, L.; Jia, R.; Jia, C.S. **Effects of different methods of acupuncture and moxibustion on functional constipation in**

rats: a comparative study. Zhonqquo Zhen Jiu. 2017; 37(5): 527- 533 [PubMed]

Xu, X.; Zheng, C.; Zhang, M.; Wang, W.; Huang, G. **A randomised controlled trial of acupuncture to treat functional constipation: design and protoco l.** BMC Complement Atern Med. 2014; 14: 423 [PubMed]

Xu, Z.; Li, R.; Zhu, C.; Li, M. **Effect of acupuncture treatment for weight loss on gut flora in patients with simple obesity.** Acupunct Med. 2013; 31(1):116-7 [PubMed]

Yoon, J.Y.; Cha, J.M.; Oh, J.K.; Tan, P.L.; Kim, S.H.; Kwak, M.S.; Jeon, J.W.; Shin, H.P. **Probiotics Ameliorate Stool Consistency in Patients with Chronic Constipation: A Randomised, Double-Blind, Placebo- Controlled Study.** Dig Dis Sci. 2018. doi: 10.1007/s10620-018-5139-8. [PubMed]
Yu, Z.; Xu, B. **Clinicai Advantages and Acupoint Selection of Treatment of Chronic Functional Constipationwith Acupuncture Therapy.** Zhonqquo Zhen Jiu. 2016; 41(4): 365-8 [PubMed]

Yang, J.P.; Liu, J.Y.; Gu, H.Y.; Lv, W.L.; Zhao, H.; Li, G.P. **Randomised controlled trials of acupuncture and moxibustion for post-stroke constipation: a meta analysis.** Zhonqquo zhen Jiu. 2014; 34(8): 833-6 [PubMed]

Yang, Y.X.; He, M.; Hu, G.; Wei, J.; Pages, P.; Yang, X.H.; Bourdu-Naturel, S. **Effect of a fermented milk containing bifidobacterium lactis DN-173010 on chinese constipated women.** World Journal of Gastroenterology. Beijing. Vol. 14. Num. 40. 2008. p. 6237-6243. [PubMed]

Yeh, C.C.; Wang, C.H.; Maa, S.H. **A conceptual framework of the effectiveness of acupuncture.** Hu Li Za Zhi. 2007; 54(4): 5-9 [PubMed]

Zheng, H.; Liu, Z.S.; Zhang, W.; Chen, M.; Zhong, F.; Jing, X.H.; Rong, P.J.; Zhu, W.Z.; Wang, F.C.; Liu, Z.B.; Tang, C.Z.; Wang, S.J.; Zhou, M.Q.; Li, Y.; Zhu, B. **Acupuncture for patients with chronic functional constipation: A randomized controlled trial.** Neurogastroenterol Motil. 2018; **doi: 10.1111/nmo.13307** [PubMed].

Zhou, S.L.; Zhang, X.L.; Wang, J.H. **Comparison of electroacupuncture and medical treatment for functional constipation: a systematic review and meta-analysis.** Acupunct Med. 2017; 35(5): 324-331 [PubMed]

75

Zhou, W.; Benharash, P. **Effects and mechanisms of acupuncture based on the principle of meridians.** J Acupunct Median. 2014; 7(4): 190-3 [PubMed]

yes

I want morebooks!

Buy your books fast and straightforward online - at one of world's fastest growing online book stores! Environmentally sound due to Print-on-Demand technologies.

Buy your books online at
www.morebooks.shop

Kaufen Sie Ihre Bücher schnell und unkompliziert online – auf einer der am schnellsten wachsenden Buchhandelsplattformen weltweit! Dank Print-On-Demand umwelt- und ressourcenschonend produzi ert.

Bücher schneller online kaufen
www.morebooks.shop

Printed by Books on Demand GmbH, Norderstedt / Germany

Nelson Ivan Chavez Mostajo

SÍNDROMES GERIÁTRICAS PARA MÉDICOS RESIDENTES

Nelson Ivan Chavez Mostajo

SÍNDROMES GERIÁTRICAS PARA MÉDICOS RESIDENTES

Integrar o B-Learning em Geriatria: uma nova era no ens no

ScienciaScripts

This book is a translation from the original published under ISBN 978-613-9-40861-0.

Publisher:
Sciencia Scripts
is a trademark of
Dodo Books Indian Ocean Ltd. and OmniScriptum S.R.L publishing group

120 High Road, East Finchley, London, N2 9ED, United Kingdom
Str. Armeneasca 28/1, office 1, Chisinau MD-2012, Republic of Moldova, Europe
Printed at: see last page
ISBN: 978-620-7-84540-8

INTEGRAR O B-LEARNING NA GERIATRFA: UMA NOVA ERA NA EDUCAÇÃO

PRÓLOGO

É com prazer que apresentamos este livro intitulado "Ensino de Sintomas Geriátricos a Médicos Residentes a partir do B-Learning", que se centra na integração da aprendizagem híbrida (B-Learning) no ensino da geriatria. Este texto surge em resposta à necessidade crescente de adaptar a formação médica às exigências actuais e futuras de uma população envelhecida, oferecendo ferramentas educativas que combinam o ensino presencial com as tecnologias digitais.

A Geriatria, enquanto ramo da medicina, dedica-se ao estudo e tratamento das doenças que afectam os idosos, bem como à promoção de um envelhecimento saudável. Neste contexto, torna-se crucial um ensino eficaz e atualizado das síndromes geriátricas, de modo a formar profissionais capazes de lidar com os desafios complexos que esta fase da vida apresenta. As síndromes geriátricas, tais como a imobilidade, a sonolência, a incontinência urinária, o défice cognitivo, a sarcopenia e a fragilidade, representam condições que afectam significativamente a qualidade de vida dos adultos mais velhos e requerem uma abordagem interdisciplinar e personalizada.

Este livro não só apresenta um compêndio abrangente de conhecimentos teóricos e práticos sobre estas síndromes, como também introduz o método B-Learning como uma estratégia inovadora para a educação médica. Através desta metodologia, os médicos residentes têm a oportunidade de aprender de uma forma flexível e dinâmica, combinando as vantagens do ensino presencial com as possibilidades oferecidas pelas ferramentas digitais.O principal objetivo deste texto é fornecer aos médicos residentes um recurso abrangente e acessível que lhes permita desenvolver competências clínicas e capacidades críticas na gestão de síndromes geriátricas. Através de estudos de caso, exercícios interactivos e recursos multimédia, os leitores poderão aprofundar os aspectos clínicos e éticos dos cuidados geriátricos, fomentando uma compreensão holística e humanizada dos cuidados ao idoso.

Gostaria de expressar a minha gratidão a todos os colaboradores que tornaram este livro possível. A sua dedicação e empenho na educação médica e no bem-estar dos adultos mais velhos foram fundamentais para a criação deste livro. Espero que este livro sirva como um guia valioso e um incentivo para que os médicos residentes continuem a sua educação médica. formação com entusiasmo e dedicação, sempre em busca da excelência nos cuidados geriátricos.

Por último, convido os leitores a explorar e a tirar o máximo partido dos conteúdos e recursos apresentados neste livro, na convicção de que a aprendizagem ao longo da vida e a aprendizagem inovadora são essenciais para enfrentar os desafios do envelhecimento da população e para melhorar a qualidade de vida dos nossos idosos.

M.Sc Nelson Ivan Chavez Mostajo Cochabamba 31 de maio de 2024

CAPÍTULO 1
INTRODUÇÃO À GERIATRIA E À GERONTOLOGIA

1. Introdução à Geriatria e à Gerontologia

A geriatria e a gerontologia são disciplinas que se ocupam da prevenção, do retardamento e da modificação dos processos biológicos, psicológicos, sociais e económicos associados ao envelhecimento. A gerontologia estuda os aspectos do envelhecimento e da velhice ao longo da vida do indivíduo, abordando-os numa perspetiva biopsicossocial (Zarebski, 2021).

A gerontologia, definida pelo funcionalista americano Robert Butler, é a ciência que investiga o envelhecimento em todas as suas dimensões. Butler salientou a importância do estudo do envelhecimento humano já no século XVII, focando os contributos significativos de vários domínios científicos.

A principal diferença entre a geriatria e a gerontologia reside no facto de a geriatria ser, antes de mais, uma disciplina de cuidados, que integra técnicas, conhecimentos e competências que visam a preservação do funcionamento físico, psicológico e social ao longo da vida. A gerontologia, por outro lado, não se concentra apenas no aspeto dos cuidados, mas também investiga a evolução cronológica de todas as dimensões do ser humano.

O objetivo da geriatria é promover uma velhice ativa e saudável ao longo de todo o ciclo de vida. Esta disciplina centra-se nos problemas específicos da velhice, como a polifarmácia, a diabetes e as perturbações da memória, e na implementação de novos métodos de cuidados, como os centros de dia, os hospitais de doentes crónicos e as unidades geriátricas (Zarebski, 2021). (Zarebski, 2021).

A investigação em gerontologia engloba aspectos sociológicos, económicos, epidemiológicos e fisiológicos, com o objetivo de compreender de forma abrangente o processo de envelhecimento. Além disso, esta disciplina é fundamental para conceber e melhorar os programas de bem-estar e de cuidados para a população idosa, promovendo um envelhecimento ativo e saudável.

1.1. Geriatria e Gerontologia: Definição e cobertura

As acções em medicina geriátrica são abrangentes e holísticas, considerando o ambiente, o estado mental, a co-morbilidade, o grau de vida independente e social, bem como os pontos de vista do sujeito e dos seus cuidadores não profissionais. O objetivo final é conseguir que os adultos com mais de 65 anos estejam livres de doenças crónicas e desfrutem de uma boa qualidade de vida. (Mazzini et al., 2023).

O objetivo geral da geriatria é a prevenção da doença através da promoção da saúde e da proteção da doença, do diagnóstico precoce, do tratamento integral e da gestão da patologia crónica mediante o controlo dos factores de risco. Além disso, procura evitar consequências negativas para a saúde quando a doença já se manifestou (Gonzalez-Montalvo et al., 2020).

3

No domínio da saúde pública, a Geriatria e a Gerontologia centram-se no estudo e no cuidado dos idosos como componentes vitais da sociedade. A partir desta abordagem, o objetivo é melhorar a qualidade de vida dos adultos mais velhos e prevenir as doenças associadas ao envelhecimento (Pifia et al., 2022).É importante notar que o aumento da população idosa levou a mudanças significativas no perfil da doença. A maioria das doenças crónicas limita hoje a capacidade funcional, enquanto as condições específicas da velhice, como a demência, o delirium, a diabetes, a caquexia, a desnutrição e as úlceras de pressão, são responsáveis pela maior incapacidade funcional nos idosos. As doenças crónicas, muitas das quais são influenciadas pelo estilo de vida, representam até 90% das principais causas de morte prematura e de doença, com a presença de doenças degenerativas a aumentar também. O processo de envelhecimento é uma fase que todos os indivíduos experimentam e que muitas vezes limita ou complica o desempenho das actividades diárias (Burgos et al., 2023).

1.2. Importância e relevância na sociedade atual

A evolução progressiva das estruturas familiares afecta as formas de assistência às pessoas idosas, manifestando-se em diversas formas de discriminação, tanto implícitas como explícitas. Isto exige que os profissionais assumam um papel ativo. A evolução para o que se designa por "nova família" ou "família poliédrica", caracterizada pela redução do número de membros, da taxa de natalidade, da duração da coabitação intergeracional, da dimensão das habitações e das mudanças no papel e localização das mulheres, coloca desafios dinâmicos nos cuidados geriátricos, criando exigências específicas aos profissionais de saúde. (Parrales & Molina, 2020) A "geriatrização" dos recursos, benefícios e serviços tornou-se uma realidade palpável, não só no domínio dos cuidados de saúde, mas também alargada aos cuidados aos idosos em todos os sectores da atividade social (Landazabal & Barboza, 2020). (Landazabal & Barboza, 2020) Neste contexto, os profissionais não só prestam cuidados aos pacientes, como também devem facilitar e desenvolver programas de promoção, prevenção, reabilitação e assistência social, gerir eficazmente as necessidades, ter em conta os desejos dos interessados, facilitar a participação da família, conhecer as alternativas residenciais e gerir os recursos económicos.

Além disso, o número crescente de pessoas idosas, que chegam à velhice em condições físicas e psicológicas heterogéneas, aumenta a procura de uma maior qualidade de cuidados. O processo de envelhecimento da população gerou um aumento exponencial da população geriátrica, afectando prioritariamente a pirâmide populacional e gerando exigências sociais e assistenciais muito específicas, como os problemas de socialização e a necessidade de novas prestações e serviços especializados. Entre os factores do lado da procura, é amplamente reconhecida a necessidade de conhecimentos em gerontologia e geriatria entre os profissionais que trabalham em cuidados sociais e/ou de saúde. (Arias et al., 2020) A compreensão e a adaptação dos serviços às necessidades específicas dos idosos são fundamentais para garantir um envelhecimento ativo e saudável. (Zubiria, 2024; Malan Valente, 2024).

1.3. Objectivos e vantagens de estudar Geriatria e Gerontologia

Os médicos e enfermeiros dos cuidados de saúde primários devem adquirir conhecimentos específicos que lhes permitam prestar aos idosos da sua carteira de serviços cuidados de qualidade adequados, planear programas de promoção da saúde, prevenir e evitar complicações, detetar precocemente e tratar as doenças da velhice, e incluir hábitos e atitudes saudáveis. Os especialistas em geriatria, bem como os internistas com formação geriátrica, devem receber uma formação abrangente que lhes permita gerir as situações clinicamente complexas e socialmente desprotegidas dos doentes idosos. Isto inclui a aplicação de princípios de funcionalidade, participação e independência, e a assunção de um papel essencial nos processos de avaliação, prevenção e tratamento. Além disso, devem contribuir, no âmbito de uma equipa multidisciplinar, para o tratamento de grupos específicos de doentes em serviços de grande complexidade que exijam hospitalização ou cuidados sociais gerontológicos. Os residentes de Geriatria e Gerontologia, depois de se familiarizarem com os conceitos básicos do envelhecimento, desenvolverão capacidades de pensamento crítico e compreenderão a complexidade multifatorial do processo de envelhecimento. Estes conhecimentos permitir-lhes-ão lidar eficazmente com os cuidados prestados aos doentes idosos, descobrindo os aspectos atractivos e desafiantes desta especialidade.

Estes profissionais terão também um conhecimento aprofundado das características geriátricas dos meios de diagnóstico, terapêuticos e preventivos partilhados com o resto da medicina clínica. Participar no desenvolvimento da medicina especializada e multidisciplinar do idoso. Aprenderão a aplicar estes conhecimentos em intervenções clínicas e programas de saúde específicos, e desenvolverão a capacidade de identificar, conceber e aplicar projectos de investigação que ajudem a resolver a falta de evidência científica no campo geriátrico, centrando-se nos aspectos preventivos e terapêuticos de determinadas doenças.

2. Aspectos biológicos do envelhecimento

O processo de envelhecimento deve ser entendido como um fenómeno natural, progressivo, deletério e universal, uma vez que afecta todos os sistemas do corpo. Biologicamente, observam-se alterações no ADN, na expressão dos genes e na função celular. Além disso, há alterações no sistema imunitário e na homeostase do organismo. É fundamental ter em conta vários aspectos biológicos que influenciam este processo, como a acumulação de danos celulares e a diminuição da capacidade regenerativa dos tecidos, que são factores chave no envelhecimento (Ramfrez, 2023). (Ramfrez, 2023).

A inflamação crónica e o stress oxidativo também desempenham um papel importante neste processo. Além disso, a imunossenescência, ou envelhecimento do sistema imunitário, aumenta a vulnerabilidade a várias doenças. As alterações hormonais têm igualmente impacto, contribuindo para o aparecimento de doenças relacionadas com a idade e para a diminuição da capacidade funcional. (Ramfrez, 2023)

A primeira fase do envelhecimento marca a transição do período de desenvolvimento pós-natal - em que o organismo aumenta de tamanho e de peso - para um período de estagnação em que o crescimento cessa. A partir deste momento, o tamanho dos órgãos é mantido pelo crescimento das células, que acabam por entrar na fase G0 do ciclo celular, cessando a divisão e atingindo um tamanho constante. Estas células adaptam-se às exigências funcionais através de um sistema de feedback local que controla o seu crescimento. Ao longo do tempo, as células podem aumentar de tamanho, por vezes de forma patológica, até chegarem a um ponto de estase nuclear, em que a divisão celular é interrompida devido à presença de uma quantidade maciça de marcadores sefialíticos (CISNEROS, 2021). Estas alterações incluem alterações na expressão genética e a acumulação de danos no ADN, que contribuem para a deterioração da homeostase celular e para o desenvolvimento de doenças relacionadas com o envelhecimento. (CISNEROS, 2021)

2.1. Alterações fisiológicas e anatómicas no processo de envelhecimento

O processo de envelhecimento é caracterizado por uma série de alterações fisiológicas e anatómicas que afectam vários sistemas corporais. Estas mudanças incluem alterações na cromatina, na metilação do ADN, nas modificações das histonas e nos perfis de ARN que têm profundas implicações no funcionamento celular (Kanasi et al., 2016).

No cérebro, o envelhecimento leva a alterações morfológicas na espessura cortical, na área de superfície e no volume de massa cinzenta, reflectindo alterações histológicas específicas durante o processo de envelhecimento (Lemaître et al., 2012). As alterações relacionadas com a idade na função de deglutição, influenciadas por alterações anatómicas e fisiológicas na cabeça e no pescoço, aumentam o risco de disfagia em adultos mais velhos (Ney et al., 2009). Além disso, são observados desvios significativos da linearidade na estrutura cerebral, indicativos de alterações relacionadas com a idade em áreas como o córtex cerebral e o diencéfalo (Jernigan et al., 1991).

No sistema músculo-esquelético, o envelhecimento pode levar a variações anatómicas e alterações ósseas, especialmente na região lombossacroilíaca da coluna vertebral (Scilimati et al., 2022). As alterações fisiológicas e anatómicas do pavimento pélvico durante a gravidez e o pós-parto podem afetar o suporte dos órgãos pélvicos e a continência (Gondim et al., 2022).

O envelhecimento também envolve alterações nos mecanismos homeostáticos celulares, redução da massa dos órgãos e diminuição da reserva funcional dos sistemas corporais (Nigam et al., 2012). Essas alterações são consequência da redução da atividade física e do próprio processo de envelhecimento (Amarya et al., 2018). Os doentes crónicos apresentam uma complexidade que aumenta com a idade, acumulando problemas como o comprometimento da memória, dificuldades na marcha, perda de audição e visão, aumentando a sua vulnerabilidade à doença aguda.

Na avaliação global do doente geriátrico, é essencial considerar aspectos como o estado funcional e a autonomia, o estado psicológico, emocional e motivacional, o ambiente social, a qualidade de vida e os papéis sociais, o estado socioeconómico, a integração familiar, a dependência, a adaptação e a realização social (Gonzalez-Montalvo et al., 2020). O objetivo desta avaliação é categorizar, identificar, avaliar e tomar decisões sobre as acções a empreender, incluindo o tratamento e o acompanhamento.

A população idosa é diversificada em termos de saúde e de doenças. As peculiaridades fisiopatológicas e clínico-terapêuticas em termos de resposta e evolução podem variar significativamente de um paciente idoso para outro, devido ao processo de envelhecimento, às múltiplas doenças crónicas associadas e aos seus tratamentos, bem como aos diferentes antecedentes pessoais e factores ambientais (Reyna et al., 2021).

Em conclusão, o envelhecimento manifesta-se através de uma multiplicidade de alterações fisiológicas e anatómicas em vários sistemas do corpo. A compreensão destas alterações é crucial para abordar os problemas de saúde relacionados com a idade e desenvolver intervenções específicas para promover um envelhecimento saudável.

2.2. Teorias biológicas e não biológicas do envelhecimento

Considerados como teóricos mecânicos (conceção cartesiana), assumem o envelhecimento como um processo que conduz inexoravelmente à deterioração e à perda de função, apresentando uma visão dos idosos como indivíduos fracos e passivos. (Yungplut, 2024).

No entanto, o envelhecimento é hoje concebido com uma nuance mais dinâmica e complexa, envolvendo muitos factores, designados por teorias do envelhecimento biológico.

Seguindo os critérios de Aristóteles, só no século XIX é que o envelhecimento foi considerado como um estudo possível. Desde o início do conhecimento biológico e ao longo da Idade Média e da Idade Moderna, considerava-se que o envelhecimento ocorria como consequência do calor gerado no corpo e que acabava por ser consumido. Foi a partir do início do século XIX, depois de séculos a considerar que o homem quase não mudava com o passar dos anos, e que era muito fácil ser um velho sábio, que começaram a ser avançadas muitas hipóteses e hipóteses. investigações relacionadas com o crescimento embrionário, o desenvolvimento e o envelhecimento (Garcfa Montes de Oca & en Estomatologfa).

As teorias do envelhecimento humano são classificadas em dois grupos, consoante a disciplina que as gerou: biológicas e extrabiológicas. Historicamente, a biologia explicou o envelhecimento com base nas mudanças que ocorrem no organismo ao longo dos anos. No entanto, a partir de meados do século XX, o carácter multicultural da sociedade e o crescente interesse pelos adultos mais velhos demonstraram a necessidade de rever a conceção do envelhecimento a partir de diferentes perspectivas, como a epidemiologia, a psicologia e a sociologia. (LA)(Grande Aranda, 2024).

2.3. Prevenção e gestão de doenças comuns na velhice

A prevenção de doenças na velhice exige uma compreensão aprofundada dos mecanismos moleculares subjacentes ao envelhecimento e às patologias relacionadas com a idade. Um fator-chave é a inflamação crónica, caracterizada pela regulação positiva de mediadores pró-inflamatórios devido a um desequilíbrio redox, que desempenha um papel importante no envelhecimento e nas doenças associadas (Chung et al., 2009). Além disso, a inflamação crónica de baixo grau, perpetuada por factores dietéticos, pode promover condições como a obesidade e a osteoporose, realçando a importância das intervenções no estilo de vida para a prevenção de doenças (Ilich et al., 2014).

No domínio da saúde renal, uma abordagem abrangente que inclua a deteção precoce e o tratamento atempado é crucial para atenuar a insuficiência renal e prevenir a doença renal crónica, salientando a lentidão e a progressão desta patologia (Carrillo-Ucafiay et al., 2022). As estratégias preventivas devem, por conseguinte, abordar os processos inflamatórios subjacentes, as influências dietéticas e as vulnerabilidades específicas dos órgãos, a fim de promover um envelhecimento saudável e reduzir o peso das doenças relacionadas com a idade.

A Geriatria é atualmente reconhecida como uma ciência com identidade própria, integrando as suas próprias disciplinas e procurando a integração com as ciências básicas e clínicas, em vez de ser apenas uma síntese de conhecimentos isolados. Inicialmente, estes conhecimentos eram essenciais para os médicos tradicionais e para a sobrevivência dos doentes crónicos com mau prognóstico. No contexto dos doentes terminais, o objetivo tem sido manter o doente no seu ambiente familiar, tentando gerir as crises sem necessariamente as resolver. Além disso, é crucial reconhecer as doenças que afectam mais frequentemente os idosos, como a hipertensão, que afecta 50% da população com mais de 60 anos. As principais causas destas doenças são as doenças cardiovasculares, as artropatias (15-30%), a artrite reumatoide (10%) e a osteoartrose das grandes articulações (Montesino et al., 2022; Álvarez-Ochoa et al., 2022).

O geriatra deve ser capaz de filtrar e organizar uma grande quantidade de informações desorganizadas, adquirindo habilmente pequenas doses de tecnologia que beneficiarão os pacientes. O seu trabalho baseia-se na premissa de "curar em casos particulares, aliviar frequentemente e confortar sempre", seguindo a famosa frase de Francesc de Paula

3.Aspectos psicológicos e sociais do envelhecimento

O envelhecimento dos adultos mais velhos implica uma série de mudanças psicológicas que envolvem a aceitação deste processo. Estas mudanças podem gerar conflitos internos e o medo de se tornarem um fardo para os outros, aspectos profundamente sentidos por muitas pessoas idosas. (Huerta Pozo, 2021).

(Agualongo Chela & Ninabanda Agualongo, 2023) Estes medos provocam reacções nas pessoas idosas que, embora semelhantes às experimentadas em fases anteriores da vida, dificultam a sua rápida adaptação ao ambiente atual.

À medida que envelhecemos, as nossas relações pessoais sofrem transformações significativas. O processo de envelhecimento coloca-nos num ambiente sócio-afetivo diferente daquele a que estamos habituados, afectando o nosso papel na sociedade. As mudanças psicológicas que acompanham o envelhecimento, juntamente com as características individuais de cada pessoa, têm impacto em todas as funções sociais em que participamos, incluindo os sistemas sociais de coping que nos dão apoio em momentos de aflição ou que podem mesmo ser utilizados para obter ganhos económicos.

Alvear (2024) salienta que o homem, ao nascer e ao morrer, é um ser social. Quando chegamos à velhice, vivemos o fenómeno do envelhecimento, que implica mudanças tanto na biologia como na memória. Estas alterações desestruturam a essência do que somos e afectam a nossa atitude global, criando uma lacuna interna e afectando a nossa esfera psico-emocional.

Com o envelhecimento, a pessoa idosa vai perdendo gradualmente as qualidades que a definem aos olhos do próprio e dos outros. Esta situação pode conduzir ao isolamento e a sentimentos de inadaptação ao meio envolvente. À medida que envelhecemos, as nossas interacções com o ambiente podem tornar-se cada vez mais complexas. O ambiente familiar e social tende a diminuir, exacerbando estes sentimentos de alienação (Tomala & Rivera, 2023).

3.1. Saúde mental e emocional na velhice

O envelhecimento tende a ser mal interpretado como uma deterioração física, emocional e cognitiva. (Iglesias et al.2024) Com o passar dos anos, é verdade que as pessoas sofrem alterações emocionais e cognitivas, mas nem todas são negativas.

Os estudos mostram que as cinco dimensões emocionais (stress, emoções positivas e negativas, ansiedade e depressão) diminuem significativamente com a idade. (Camacho Torres & Trejo Bravo, 2023) Tanto a ansiedade como a depressão diminuem significativamente após os 65 anos de idade. Isto deve-se em grande parte ao declínio cognitivo associado ao envelhecimento, bem como a doenças crónicas e a acontecimentos de vida stressantes. Uma das principais conclusões sobre os adultos mais velhos (mais de 65 anos) é que cerca de metade deles pode precisar de apoio médico para melhorar a sua qualidade de vida, enquanto os restantes aprendem a tolerar os seus sintomas e a continuar com as suas actividades diárias. Este fenómeno pode ser explicado por uma abordagem analítica para lidar com os obstáculos que surgem durante a velhice, bem como pela adoção de um pensamento positivo ou resiliente. (Velasco Gaibor, 2022)(Salas et al.2021).

Alguns elementos salientes que podem contribuir para uma velhice "bem sucedida" incluem: manter uma boa saúde, participar em actividades gratificantes e produtivas, utilizar o tempo de forma criativa, ter a capacidade de se adaptar a circunstâncias mutáveis e stressantes, enfrentar os contratempos da vida com bom humor, ter expectativas realistas, estimular a atividade cognitiva através da educação contínua e do interesse por assuntos actuais, manter uma vida social ativa, ter uma boa situação económica ou sentir-se satisfeito com o que tem.

3.2. Adaptação à mudança de papéis e circunstâncias

Tornar-se avô após os 65 anos implica muitas vezes a necessidade de modificar os padrões de relacionamento com os filhos, respeitando o novo papel dos seus parceiros.

Esta mudança na dinâmica familiar coincide frequentemente com a reforma, o que permite ao avô assumir um novo papel na relação com os netos, alterando parcialmente a dinâmica geracional. Este papel pode ser muito gratificante e compensa, em certa medida, a saída do mundo do trabalho. Por conseguinte, a reforma não representa necessariamente uma diminuição do número de papéis sociais, mas sim um período de ajustamento e de recombinação de papéis previamente estabelecidos. Para os casais mais velhos, a reforma significa também uma adaptação a uma coabitação diária e total mais longa. Com efeito, contrariamente ao tempo passado em conjunto durante o trabalho e os fins-de-semana, a reforma implica um aumento significativo do tempo passado em conjunto. Além disso, enfrentam o desafio de se integrarem no grupo de pessoas mais velhas, o que pode significar uma nova dinâmica social para eles.

Nas situações em que um dos parceiros fica viúvo, é necessário adaptar-se a mudanças significativas na sua situação socio-afectiva. A viuvez pode complicar a aceitação desta nova realidade e antecipa desafios na gestão adequada das doenças, necessárias para preservar uma saúde óptima. A adaptação à viuvez implica, por conseguinte, a passagem por múltiplas variáveis que podem dificultar a transição para uma nova fase da vida.

3.3. Apoio social e redes de apoio

No âmbito da teoria do stress, é de extrema importância ter em conta dois conceitos essenciais: a perceção do apoio social e a socialização relacional (Cedillo, 2020) (Olivares Ramfrez, 2021). Se partirmos da premissa de que a falta de apoio pode gerar stress, podemos afirmar que o que realmente importa é, por um lado, a relevância da perceção subjectiva de cada indivíduo numa situação específica, na qual se sente socialmente apoiado.

Por outro lado, o sistema de socialização pessoal é também relevante. Nesta perspetiva, o ser humano tem a capacidade de organizar o seu ambiente interpessoal de forma estruturada, em função da presença ou ausência de pessoas significativas, utilizando variáveis como o número, a qualidade e a estabilidade.

Os organismos sociais que exercem influência sobre a pessoa formam uma rede pessoal, que pode ser dividida entre os organismos significativos que têm uma influência direta sobre o indivíduo que recebe a rede e aqueles a cujo "papel" o sujeito adere (rede de relações). (Reyes Sanchez & Sandoval Bocanegra, 2023).

O apoio em momentos difíceis e o sentimento de fazer parte de uma rede de relações gratificantes que proporcionam ajuda, apreço, afeto e informação adequada podem ter efeitos protectores significativos na saúde. As relações sociais podem proporcionar ao indivíduo um sentimento de pertença, identidade, objetivo e orientação com base nos papéis assumidos no grupo. Inversamente, a ausência de vínculos significativos pode desencadear processos de ampla reatividade, tornando o indivíduo vulnerável a exigências de auto-atribuição de culpa pela sua situação e às pressões das normas sociais, o que altera a autoestima e gera sentimentos de isolamento (Reyes Sanchez & Sandoval Bocanegra, 2023).

Neste contexto, face a ameaças de perda ou quando confrontados com uma norma, a presença de relações significativas desempenha um papel crucial através de processos de regulação, proteção ou restrição. Isto facilita a adaptação a novas exigências e desafios, demonstrando como as ligações interpessoais significativas não só enriquecem a nossa vida emocional, como também são fundamentais para a nossa capacidade de adaptação e bem-estar psicológico.

4. Aspectos éticos e jurídicos em geriatria e gerontologia

A população idosa é particularmente sensível e exige considerações éticas que tenham em conta a sua vulnerabilidade. Embora os princípios bioéticos clássicos sejam geralmente aplicáveis, os princípios da autonomia, da beneficência e da não-maleficência revestem-se de especial importância para os idosos.

Uma vez que os idosos apresentam frequentemente uma elevada prevalência de patologias múltiplas e, nalguns casos, limitações funcionais, os cuidados médicos são preferencialmente orientados para o conforto e o alívio, em vez de intervenções diagnósticas e terapêuticas agressivas. Este facto ajuda a evitar a institucionalização e favorece a manutenção da autonomia do doente.

É fundamental que as intervenções médicas respeitem os desejos e a autonomia do doente. É também vital assegurar que os princípios da beneficência e da não maleficência sejam respeitados, proporcionando cuidados abrangentes e de qualidade que promovam a autonomia do doente na tomada de decisões sobre a sua saúde.

No caso dos idosos hospitalizados, é importante planear adequadamente a admissão, evitando acções urgentes desnecessárias e o uso excessivo de exames complementares que não modificam o plano de cuidados. Além disso, a tecnologia e os avanços médicos não devem substituir a importância de uma história detalhada e de um exame físico, nem minimizar a relevância de manter um relatório abrangente que reflicta os desejos e as circunstâncias do doente.

O consentimento informado deve ser obtido para cada procedimento de diagnóstico e terapêutico que envolva risco, mesmo para os procedimentos de menor importância. Este consentimento deve ser documentado no registo médico, juntamente com as medidas tomadas para apoiar o paciente na tomada de decisões relacionadas. No caso de o paciente assinar um consentimento informado ou de ser internado num estabelecimento de saúde, é essencial identificar claramente os interlocutores válidos nas situações em que o paciente não pode exprimir a sua vontade. Esta informação deve ser registada no processo clínico, respeitando a legislação aplicável, e deve incluir as decisões anteriores tomadas em circunstâncias clínicas específicas, explicando os valores e os princípios aceites pelo doente ao doente, à sua família ou ao seu representante legal, a fim de lhe dar o apoio necessário para tomar decisões adequadas.

4.1. Princípios éticos na prestação de cuidados aos idosos

Outro princípio importante da ética médica é o da "não maleficência", ou seja, a obrigação de não causar danos. Este princípio deve ser respeitado tanto antes de diagnosticar uma doença como antes de propor um tratamento. Ligado a este princípio está o princípio da "beneficência", ou seja, promover o bem do doente, propondo e aplicando tratamentos para o bem do doente e causando-lhe o menor dano possível.

Para além disso, o médico é obrigado a respeitar e manter a confidencialidade de toda a informação que lhe chega através da sua prática, assumindo o sigilo profissional.

Esta obrigação é adquirida por várias razões. O indivíduo dirige-se ao médico com a confiança de que as informações que fornece para tratamento não serão divulgadas, e o médico não pode realizar o seu trabalho de forma eficaz sem informações verdadeiras e completas, que só podem ser obtidas através da dissipação dos receios do doente. Por conseguinte, o médico deve manter a confidencialidade sobre o que o doente lhe diz, a menos que esteja convencido de que a não divulgação da informação resultaria em prejuízo para o indivíduo ou para a sociedade.

O Código de Ética Médica estabelece que os médicos devem respeitar a dignidade da pessoa e a sua privacidade, tanto dos idosos como das outras pessoas, evitando a discriminação em função da idade. Assim, a única alternativa terapêutica para a pessoa idosa não deve ser a que deriva da sua idade, mas deve ter em conta o juízo clínico e atuar com iguais qualidades éticas.

Ligado à dignidade do indivíduo está o princípio da autonomia, ou seja, o doente tem o direito de receber a informação necessária para tomar as suas próprias decisões relativamente à sua vida e à sua saúde.

4.2. Legislação e direitos dos idosos

O direito humanitário internacional foi especificamente concebido para proteger as pessoas idosas. Em 1991, a Assembleia Geral das Nações Unidas realizou na sua sede em Viena a Assembleia Mundial sobre o Envelhecimento: "Rumo a uma sociedade para todas as idades". Reafirma o direito humano a uma vida longa e saudável, à proteção jurídica, à proteção física e ao apoio na velhice, considerando este objetivo a nível nacional e internacional, nos domínios social, económico e humanitário. (Torres et al.2023).

A Carta dos Direitos Humanos em Gerontologia é um documento sócio-político que pretende explicar, analisar e denunciar o complexo e variado fenómeno do envelhecimento em termos de direitos humanos. Pretende também chamar a atenção para a realidade existente, as suas raízes e o sistema socioeconómico que lhe está subjacente como causa imediata de outras realidades e problemas que muitas vezes escondem os anteriores.

E, em particular, visa empreender esforços, acções e denúncias com o objetivo de promover uma mudança social e cultural entre as pessoas envolvidas de diferentes formas: profissionais, familiares, utentes e os próprios idosos. (Torres et al.2023).

A legislação relacionada ccm a velhice, a nível internacional, nacional e regional, é atualmente muito extensa. Este facto deve-se ao aumento da esperança de vida e ao crescimento da população idosa nos países ocidentais.

Entre os cinquenta países do continente africano e intertropical, foi elaborado um "Plano de Ação de Lagos", que aborda sete questões relacionadas com os reformados.

Por seu lado, a Organização de Cooperação e de Desenvolvimento Económicos implementou a Declaração de Princípios e Recomendações Distributivas relativas à provisão privada para a velhice nos vinte países mais industrializados.

Na União Europeia, o Tratado de Maastricht considera essencial a análise das despesas sociais para identificar as situações que excedem as capacidades de um Estado e que, por outro lado, podem ser financiadas coletivamente.

4.3. Ética na investigação geriátrica

O consentimento informado é de importância vital para a utilização dos dados obtidos num protocolo de investigação. O doente geriátrico deve receber informações completas sobre os objectivos do estudo, a metodologia, as possíveis alternativas de tratamento do estudo (incluindo os cuidados habituais), as interacções medicamentosas, os benefícios e os riscos potenciais.

Também deve ser mencionado o que torna o procedimento difícil de entender, como o número de medicamentos e exames necessários. Deve ser dada ao doente a oportunidade de refletir sobre esta informação, por escrito ou verbalmente.

O investigador deve assegurar-se de que o idoso confia no profissional de saúde que conduz o protocolo e não discriminar os que não participam, premiando os que participam.Ao longo da história, houve situações em que a investigação comprometeu o bem-estar psicofísico do doente geriátrico que fazia parte do estudo. Alguns autores descrevem-nas de uma forma generalizada. Por exemplo, Hipócrates menciona os médicos abastados que efectuavam investigações inconscientes e sem escrúpulos, em nome do prestígio e do poder. Na Ética a Nicómaco, Aristóteles refere que as coisas são julgadas boas ou más em função dos seus efeitos. Galeno afirma que a funcionalidade do paciente não deve ser comprometida em nome da experimentação. E já na época contemporânea, a infame cadeia de Nuremberge assinala o marco do respeito e da consideração pelo doente que participa num estudo.

Hoje em dia, a participação de adultos mais velhos em projectos de investigação tornou-se crucial na comunidade científica, porque uma parte significativa dos avanços que a medicina alcançou no passado foi graças às numerosas contribuições e à participação voluntária de pessoas com 65 anos ou mais.

5. Avaliação geriátrica abrangente

A avaliação geriátrica abrangente (AGA) é uma ferramenta fundamental no cuidado dos adultos mais velhos. Apesar da importância da VIG, tem havido uma falta de aplicações móveis focadas nesta área, apesar do seu potencial para reduzir as hospitalizações, a incapacidade e a mortalidade (Bautista-Mier et al., 2021). A VIG é uma abordagem multidimensional que engloba aspectos médicos, afectivos, cognitivos, funcionais, sociais e espirituais, o que a torna uma ferramenta abrangente para o cuidado da população adulta mais velha (Martfnez et al., 2022). A avaliação geriátrica abrangente é de extrema importância nos cuidados e tratamento da população idosa. Ao contrário de outras fases da vida, na velhice as pessoas enfrentam frequentemente doenças crónicas e têm menos reservas fisiológicas.

Estes indivíduos têm muitas vezes dificuldade em tolerar alterações do seu equilíbrio interno, o que conduz frequentemente a perturbações funcionais secundárias à doença e, por fim, à morte. É essencial que a medicina para os idosos seja capaz de prestar cuidados adaptados a estas características específicas.

5.1. Componentes da avaliação geriátrica

A avaliação geriátrica global é uma análise rigorosa do estado de saúde do idoso para planear cuidados e tratamentos individualizados.

É também conhecida como avaliação geriátrica global e responde à necessidade de desenvolver programas de avaliação para os idosos, centrados em dois objectivos: detetar e tratar perturbações agudas ou crónicas nos idosos, a fim de reduzir os sintomas e minimizar as consequências da doença; e, em segundo lugar, identificar e intervir sobre os défices específicos que contribuem para a incapacidade e a dependência. Segundo Kirkwood e Melzer, a avaliação do idoso passa pelo conhecimento e consideração de características ligadas a quatro domínios distintos mas interdependentes: biológico, fisiológico, patológico e social. Cada um dos componentes que constituem um indivíduo, isoladamente, tem características diferentes consoante a fase de vida em que se encontra. No entanto, estas características, já de si diferentes, assumem um valor especial numa fase em que não existem tantos "recursos" disponíveis, do ponto de vista funcional, fisiológico e mesmo patológico, para levar uma vida tão autónoma quanto possível.

Os avanços simultâneos da Geriatria e da Gerontologia no final do século XX levaram a um repensar desta situação. A elevada prevalência e a influência dos processos patológicos (nomeadamente as demências e as doenças do aparelho locomotor) na incapacidade funcional, ou a atividade patológica subjacente às chamadas síndromes geriátricas (café, incontinência e imobilização), aumentam a complexidade desta já complicada etapa da vida.

5.2. Instrumentos e escalas de avaliação em geriatria

No domínio da geriatria, a avaliação geriátrica global (AGG) desempenha um papel crucial na avaliação da saúde e do bem-estar dos idosos. A AGA é uma ferramenta multidimensional que considera não só os aspectos médicos, mas também as dimensões emocional, cognitiva, funcional, social e espiritual de um indivíduo, o que a torna um instrumento fundamental para prestar cuidados holísticos aos idosos (Martfnez et al.,

2022). Esta abordagem holística é essencial para responder às diversas necessidades dos idosos, especialmente no contexto da gestão de doenças terminais, em que factores para além da própria doença podem afetar significativamente os resultados (Beracasa et al., 2021).

Embora a VGI seja uma ferramenta valiosa, a sua implementação pode, por vezes, ser um desafio devido a restrições de tempo e à falta de métodos normalizados para avaliações geriátricas em ambientes ambulatórios (Beracasa et al., 2021). Nestes casos, os instrumentos de avaliação mais curtos tornam-se essenciais para prever e gerir eficazmente os tratamentos nos idosos. Estas ferramentas mais curtas podem fornecer informações valiosas sobre o desempenho funcional, o estado cognitivo e o apoio social, o que ajuda nos processos de tomada de decisão (Beracasa et al., 2021). No domínio dos cuidados geriátricos, a utilização de aplicações móveis para uma avaliação geriátrica abrangente está a ganhar atenção pelo seu potencial para reduzir as hospitalizações, a institucionalização, a incapacidade e a mortalidade (Bautista-Mier et al., 2021).

Embora muitas aplicações móveis em geriatria se centrem na gestão de doenças crónicas, na estimulação cognitiva e na atividade física, há um reconhecimento crescente dos benefícios da incorporação de aplicações centradas nas VGI na prática clínica. Estas aplicações têm o potencial de revolucionar os cuidados geriátricos, fornecendo meios acessíveis e eficientes para efetuar avaliações e melhorar os resultados dos doentes (Bautista-Mier et al., 2021).

A avaliação funcional é um componente crítico dos cuidados geriátricos, especialmente para identificar limitações e prever a incapacidade em adultos mais velhos (Ocampo et al., 2017). A implementação de ferramentas de avaliação funcional na prática clínica pode ajudar a monitorizar a progressão das incapacidades e a orientar os planos de cuidados para os idosos, em particular os que se encontram em ambientes institucionais. Ao utilizar escalas de avaliação funcional, os prestadores de cuidados de saúde podem compreender melhor as necessidades das pessoas idosas, particularmente aquelas em ambientes institucionais. necessidades em mudança dos adultos mais velhos e adaptar as intervenções para melhorar a sua qualidade de vida (Ocampo et al., 2017).A depressão é uma preocupação comum entre a população idosa e a Escala de Depressão Geriátrica (GDS) é uma ferramenta amplamente utilizada para avaliar os sintomas depressivos em adultos mais velhos (Almeida e Almeida, 1999). A fiabilidade e a validade da GDS fazem dela um instrumento valioso para identificar e gerir a depressão em doentes geriátricos. Dada a elevada prevalência da depressão nos idosos e o seu impacto no bem-estar geral, instrumentos como a GDS desempenham um papel crucial na garantia de cuidados abrangentes para os adultos mais velhos (Almeida e Almeida, 1999).

No contexto da avaliação da dor, as escalas de classificação numérica (NRS) são normalmente utilizadas para avaliar a intensidade da dor em contextos clínicos e de investigação (Hidalgo et al., 2021). A NRS permite que os observadores atribuam pontuações subjectivas a diferentes níveis de dor, proporcionando um método normalizado de avaliação da dor. Ao utilizar ferramentas como a NRS, os profissionais de saúde podem monitorizar e gerir eficazmente a dor nos idosos, melhorando assim a sua qualidade de vida e os resultados globais dos cuidados (Hidalgo et al., 2021).

Nos serviços geriátricos espanhóis, são utilizados vários instrumentos de avaliação para avaliar a saúde e o estado funcional dos idosos (Ruano et al., 2014). Estes instrumentos são essenciais para prestar cuidados integrais adaptados às necessidades específicas dos idosos. Ao incorporar uma variedade de instrumentos de avaliação, os profissionais de saúde nos serviços geriátricos espanhóis podem garantir uma avaliação abrangente dos pacientes mais velhos, levando a intervenções mais personalizadas e eficazes (Ruano et al., 2014).

A avaliação de enfermagem em ambientes geriátricos desempenha um papel vital na captação das necessidades holísticas dos adultos mais velhos, particularmente daqueles que residem em lares de idosos (Sanchez et al., 2007). Ao utilizar instrumentos de avaliação de enfermagem abrangentes que englobam vários domínios, como os aspectos físicos, mentais, sociais e funcionais, os enfermeiros podem desenvolver planos de cuidados individualizados que respondam às necessidades específicas dos residentes mais velhos. Estes instrumentos de avaliação servem como um roteiro para a prestação de cuidados de qualidade e para a promoção do bem-estar dos idosos em ambientes institucionais (Sanchez et al., 2007).

No contexto da avaliação do risco de úlceras de pressão, a utilização de escalas padronizadas para avaliar o risco de desenvolvimento de úlceras de pressão é crucial para a prevenção destas condições debilitantes nos idosos (Fernandez et al., 2008). Estas escalas incluem a escala de Braden e a escala de Norton. Ao utilizar instrumentos de avaliação validados, é possível identificar as pessoas com risco acrescido de desenvolver úlceras de pressão e implementar intervenções específicas para mitigar esse risco. Estas escalas são instrumentos valiosos nos cuidados geriátricos, uma vez que ajudam na deteção precoce e na prevenção de lesões cutâneas em idosos vulneráveis (Fernandez et al., 2008).

O conceito de avaliação geriátrica global (AGC) é essencial tanto no contexto hospitalar como nos cuidados primários, uma vez que permite aos profissionais de saúde prestar cuidados holísticos e de elevada qualidade aos idosos (Wanden-Berghe, 2021). O pessoal médico pode obter uma compreensão holística do estado de saúde, das capacidades funcionais e dos sistemas de apoio social de um doente idoso, permitindo intervenções personalizadas que respondam às necessidades individuais dos idosos. A abordagem holística garante que os idosos recebam cuidados personalizados e eficazes que tenham em conta as suas circunstâncias e desafios únicos (Wanden-Berghe, 2021). Uma avaliação exaustiva que aborde os diferentes sistemas do doente (circulatório, respiratório, motor, sensorial, psíquico) através de uma exploração funcional permitirá conhecer o estado atual do doente e ajudar-nos-á a estabelecer as situações de saúde determinantes. Algumas doenças podem dar origem a síndromes geriátricos como a instabilidade, as síncopes, os codacidentes, as doenças iatrogénicas, a deterioração funcional e a perda de autonomia. Estas doenças podem desencadear os motivos de consulta mais comuns neste grupo etário.

Deve ser identificada a doença anterior, as doenças da história clínica que estão relacionadas com o estado atual, a localização e o tipo. Também deve ser feita uma avaliação dos antecedentes pessoais e psicossociais. É importante excluir doenças recentes, infecções, desnutrição, fracturas, doenças do sistema nervoso e sequelas de acidentes, bem como casos de maus tratos. Ao contrário do adulto jovem, o adulto idoso tem um prognóstico a curto prazo e a perda de função diminui rapidamente a qualidade de vida. A equipa multidisciplinar, constituída por médicos, enfermeiros, fisioterapeutas, terapeutas ocupacionais, psicólogos e assistentes sociais, deve procurar obter o máximo de informação na avaliação e planeamento da intervenção com o idoso. É essencial a utilização de escalas de avaliação adequadas ao contexto em que vão ser utilizadas: seja em contexto extra-hospitalar, hospitalar, residencial, domiciliário ou de cuidados de dia. O tempo necessário para completar a escala deve ser adequado ao contexto em que é aplicada.

5.3. Interpretação dos resultados e planeamento dos cuidados

Como recurso para identificar alterações na pessoa idosa, os profissionais dispõem de uma variedade de escalas ou critérios que nos aproximam da sintomatologia.

Podemos citar o Índice de Massa Corporal (IMC), a escala visual de Snellen, a escala de Folstein, a escala de depressão geriátrica, o MINIMENTAL, entre outros. Todos eles devem ser analisados com uma atitude crítica, sem se deixar levar por uma única informação. É essencial compilar uma história clínica detalhada e exacta, uma vez que esta orienta as nossas intervenções para as necessidades do indivíduo idoso. Em caso de falta de dados pessoais, procuramos sempre os familiares para preencher a lacuna.

Consideramos como parte do envelhecimento normal (dentro dos limites das perdas patológicas evitáveis) a diminuição do peso corporal, da visão ou da audição (até certo ponto); a alteração da mobilidade, os sentimentos de solidão ou de insatisfação; a diminuição do desejo sexual. Sob o termo envelhecimento normal, referimo-nos ao declínio funcional que é de esperar ao longo dos anos; consideramo-lo geralmente irreversível e não relacionado com a doença. O que é importante é encontrar o ponto exato para identificar a situação patológica e estabelecer como objetivo a manutenção e a recuperação do nível mais elevado possível de bem-estar do indivíduo.

A normalidade reside no total equilíbrio e bem-estar do ser humano. Com o passar dos anos, o indivíduo experimenta uma diminuição das suas capacidades e muitos destes sinais pré-patológicos são percepcionados como normais no processo de envelhecimento (rugas na pele, queda de cabelo, entre outros).

6. Intervenções e tratamentos em Geriatria

No domínio da saúde, o controlo dos factores de risco é crucial, uma vez que a doença é uma situação irreversível e prejudicial que pode, no entanto, ser passível de prevenção e vigilância, especialmente nas suas fases iniciais. A tónica é colocada principalmente na prevenção primária, que visa evitar o aparecimento da doença, e na prevenção secundária, que visa detetar e tratar as doenças nas suas fases iniciais. Embora seja importante prevenir a doença, é também crucial considerar que os cuidados geriátricos se centram principalmente no alívio dos sintomas para melhorar o bem-estar do doente. Por outras palavras, o objetivo é aliviar os problemas quando seria mais eficaz adotar medidas preventivas, uma vez que quanto menos sintomas aparecerem, menos tratamento será necessário (Orozco et al., 2024; Pinilla et al., 2020).

(Mufioz Toledo & Ochoa Cueva, 2023) Este conceito não se centra apenas na abordagem dos problemas numa perspetiva de saúde, mas também na consideração da auto-perceção do indivíduo, das relações sociais e da integração no seu ambiente. A atenção e o cuidado ao idoso envolvem a utilização de inúmeros recursos e estão relacionados a uma ampla gama de problemas, estratégias e elementos a serem considerados (Castro Silva, 2022).

6.1. Farmacologia na velhice

A alteração do volume de distribuição é imprevisível, uma vez que tende a aumentar para a maioria dos fármacos, conduzindo a uma semi-vida mais longa.

A eliminação de muitos medicamentos é reduzida nos idosos, especialmente através da urina. A função gastrointestinal atrasada também afecta a absorção, quer devido a um trânsito mais lento (devido à diminuição do tónus do músculo liso), quer devido ao envolvimento da mucosa ou a medicamentos como os antiácidos, que diminuem a absorção dos barbitúricos e da digoxina.

A alteração da sensibilidade aos agonistas simpaticomiméticos e colinomiméticos, as alterações da função cerebral e os efeitos do stress e da doença crónica influenciam o comportamento face à droga. Esta variabilidade nas respostas aos fármacos exige um controlo regular dos mesmos e, sempre que possível, que se evite a sua utilização.

A insuficiência renal e a atrofia de numerosos órgãos são particularmente importantes. As alterações hepáticas e renais são relevantes, uma vez que o fígado e os rins são os órgãos responsáveis pelo metabolismo e eliminação da maioria dos fármacos. Geralmente, o fígado reduz o seu metabolismo dos fármacos ao longo do tempo, embora existam alguns fármacos que não alteram o seu metabolismo. No entanto, existem algumas excepções de medicamentos cujo metabolismo é afetado. A maioria dos fármacos diminui a depuração renal, o que implica um maior efeito farmacológico, uma vez que o seu nível plasmático se mantém elevado.

6.2. Reabilitação e fisioterapia para idosos

A fisioterapia actua melhorando, a partir do seu campo de ação músculo-esquelético, tanto o tónus muscular espástico como o encurtamento e a hiperlaxidez muscular, sendo recomendada naquelas doenças cujos tratamentos médicos não produzem uma melhoria significativa. No caso de doenças degenerativas como a doença de Parkinson, a fisioterapia não se limita apenas a um tratamento paliativo a ser administrado nas fases avançadas da doença. No entanto, também não é possível reduzir completamente as incapacidades nas fases iniciais com novas acções farmacológicas, pelo que é necessário abordar tanto a incapacidade de movimento através de medicação adequada como de fisioterapia apropriada. Estes esforços ajudarão a reduzir as hipóteses claras de desenvolvimento de complicações motoras e os riscos tanto para o doente como para os seus familiares. Por esta razão, é comum agir como se a medicação nunca pudesse alcançar os resultados desejados em movimentos específicos.

De uma forma geral, a estratégia consiste em adaptar a reabilitação à deficiência presente e ao estado clínico que se manifesta. Assim, para além de um tratamento geriátrico completo (controlo dos factores de risco cardiovascular, deteção e tratamento da sarcopenia, da osteoporose, etc.), acrescentaremos uma fisioterapia específica para os pacientes que dela sofrem, o que será de grande interesse para os nossos idosos, sobretudo quando há incapacidade de realizar exercícios de transferência e a cinesiterapia ativa dificilmente pode ser realizada, pelo que a fisioterapia ativa baseará o seu tratamento na inibição dos reflexos anormais, no relaxamento muscular, nas técnicas de inibição do tónus reflexo e na inibição da função patológica.

6.3. Cuidados paliativos e cuidados no fim da vida

Os cuidados paliativos vão para além da prevenção e do tratamento da dor e de outros sintomas. Promovem também a prestação de cuidados éticos e humanos aos doentes e às famílias, bem como a assistência no tratamento espiritual, tudo num ambiente digno e confortável. O objetivo dos cuidados paliativos é controlar todos os sintomas gerados por uma doença terminal, garantir a melhor qualidade de vida possível e dar apoio aos familiares e cuidadores. O doente, embora consciente do carácter irreversível da sua doença terminal, pode e deve ser envolvido em todos os aspectos dos seus planos futuros. A importância da prevenção e a gestão global da evolução clínica de qualquer doença são razões fundamentais para envolver todos os profissionais e a equipa médica nos cuidados paliativos.

No entanto, o fim da vida nem sempre é o resultado parcial ou final do tratamento de uma doença específica. Por vezes, a morte resulta de uma combinação de problemas decorrentes do tratamento médico de sintomas parciais ou de diagnósticos secundários, bem como das limitações do envelhecimento. (Arroyo et al.2022)

A gestão desta situação complexa exige os conhecimentos e as competências da abordagem paliativa. O médico, que acompanhou o doente ao longo de toda a sua vida, assiste ao desfecho final de um processo evolutivo fisiológico complexo, que se manifesta com um ritmo final de deterioração.

7.Investigação e avanços em Geriatria e Gerontologia

Concluímos esta secção destacando, em termos gerais, algumas das áreas do conhecimento em que se verificaram os avanços mais significativos em Geriatria e Gerontologia nas últimas duas décadas.

A investigação neste domínio conduziu a uma melhor compreensão e análise dos problemas de saúde específicos da população idosa, identificando e desenvolvendo intervenções mais adequadas e determinando o seu impacto na saúde e na qualidade de vida do paciente idoso. Entre as áreas que foram identificadas como as mais avançadas estão: todos os aspectos relacionados com o processo normal de envelhecimento e as doenças crónicas que afectam os idosos.

Desenvolvimento de projectos: Existem muitas organizações que desenvolvem programas de apoio à investigação nos domínios da Geriatria e da Gerontologia. Na Europa, a IAGG-Europe criou uma rede para trocar ideias e encorajar a criação de propostas conjuntas, permitindo a sua circulação entre os seus membros. Respeitando as particularidades de cada país, são igualmente relevantes as propostas a nível nacional, bem como as que se dirigem especificamente às organizações internacionais e aos seus diversos programas.

Implementação da investigação: A execução do projeto implica a necessidade de pesquisa de pacientes, recolha de dados, trabalho de campo, trabalho de laboratório, análise de amostras. Para além dos procedimentos administrativos morosos frequentemente envolvidos, existem várias entidades equipadas com infra-estruturas e métodos para facilitar a parte prática da investigação.

7.1. Investigação básica e clínica em gerontologia

A investigação básica e clínica em gerontologia é essencial para compreender melhor os processos de envelhecimento e para desenvolver tratamentos mais eficazes para os adultos mais velhos. Além disso, proporciona uma compreensão aprofundada das doenças relacionadas com a idade e ajuda a encontrar formas de as prevenir e tratar mais eficazmente.

A investigação fundamental fornece os fundamentos essenciais sobre a natureza primordial de certos processos que condicionam o próprio objeto do conhecimento.

Por outro lado, a investigação clínica emprega métodos e técnicas para desvendar respostas a questões específicas relacionadas com a doença. A falta de investigação clínica tem um impacto direto no nível insuficiente de conhecimento dos aspectos preventivos e terapêuticos da doença em geral nas pessoas idosas. Além disso, a falta de conhecimentos sobre as manifestações clínicas, a patogénese e a fisiopatologia das doenças com elevada incidência nos idosos ou sobre a utilização e os efeitos da administração de medicamentos no organismo dos idosos constituem um grande obstáculo ao desenvolvimento da disciplina.

A maior parte da investigação nas ciências básicas carece de uma orientação clínica, uma vez que o seu objetivo é estudar os processos fundamentais e universais que nos permitem conhecer as leis gerais da natureza, o que constitui a chamada ciência pura. Por outro lado, o objetivo final ou o nível aplicado da investigação clínica só é alcançado na aquisição de informações concretas que possam ser úteis na prática clínica diária ou no planeamento do tratamento.Por conseguinte, é evidente que a abordagem das necessidades dos idosos e, por conseguinte, a qualidade da informação clínica diária, depende diretamente do desenvolvimento da investigação.

7.2. Tecnologias e abordagens inovadoras nos cuidados aos idosos

No domínio da geriatria e da gerontologia, foram desenvolvidas múltiplas tecnologias e abordagens inovadoras com o objetivo de melhorar os cuidados prestados aos idosos. Um desses desenvolvimentos é a utilização de dispositivos de monitorização remota, que permitem aos profissionais de saúde acompanhar constantemente os parâmetros vitais dos doentes a partir do conforto das suas casas. Estes dispositivos incluem sensores de atividade, monitores de tensão arterial e monitores de glicose. Além disso, foram desenvolvidas aplicações móveis e plataformas digitais para promover a comunicação entre os doentes e os seus médicos, simplificando os cuidados e reduzindo a necessidade de deslocações (Acufia Valderrama, 2022) (Vega Baudrit et al., 2024) (Mufioz Zuta).

Outra abordagem inovadora reside na utilização da realidade virtual e da realidade aumentada para melhorar a reabilitação e a fisioterapia dos idosos. Estas tecnologias permitem a recriação de ambientes virtuais que facilitam a realização de exercícios e actividades físicas, estimulando a mobilidade e a funcionalidade. Em suma, as tecnologias e abordagens inovadoras nos cuidados aos idosos estão a transformar a forma como os seus cuidados são abordados e a melhorar a sua qualidade de vida (Collazo et al., 2020) (Vega Baudrit et al., 2024).

7.3. Perspectivas e desafios futuros

A geriatria e a gerontologia enfrentam uma série de desafios e têm perspectivas promissoras para o futuro. Um dos principais desafios é o envelhecimento da população e a procura crescente de cuidados médicos especializados para os idosos.

Para tal, é necessário desenvolver programas de ensino e formação em geriatria e gerontologia, a fim de garantir que os profissionais de saúde estejam preparados para responder às necessidades desta população.

Além disso, é necessário melhorar a investigação neste domínio, a fim de desenvolver abordagens inovadoras e tratamentos eficazes para os cuidados dos idosos.

As perspectivas futuras incluem a utilização de tecnologias avançadas de cuidados de saúde, como a telemedicina e a inteligência artificial, para oferecer serviços mais acessíveis e personalizados aos idosos. Espera-se também que seja dada maior ênfase à promoção da saúde e à prevenção de doenças na velhice, através de políticas públicas e programas de educação.

Em resumo, a geriatria e a gerontologia enfrentam grandes desafios, mas também têm um grande potencial para melhorar a qualidade de vida dos idosos no futuro.

REFERÊNCIAS

• Acufia Valderrama, J. A. (2022). Disefio de un sistema de monitoreo de
pacientes com doenças ambulatórias, em apoio a um estabelecimento de saúde de cuidados
gerais, categoria II 1, do ... utp.edu.pe
• Agualongo Chela, L. G. & Ninabanda Agualongo, S. M. (2023). Alterações emocionais
relacionadas à patologia; COVID 19 em adultos mais velhos.

Centro de Salud Promejoras San Camilo precinct janeiro-abril 2023.ueb.edu.ec

• Álvarez-Ochoa, R., Torres-Criollo, L. M., Ortega, J. P. G., Coronel, D. C. I.,
Cayamcela, D. M. B., Pelaez, V. D. R. L., & Salinas, A. S. S. (2022). Fatores de risco
para hipertensão em adultos. Uma revisão crítica. Revista Latinoamericana de
Hipertensi6n, 17(2). ucv.ve
• Alvear, M. E. C. (2024). Envejecimiento humano: un analisis integral desde la perspetiva
de la medicina interna. RECIAMUC. reciamuc.com
• Amarya, S., Singh, K., & Sabharwal, M. (2018). Processo de envelhecimento e
alterações fisiológicas... https://doi.org/10.5772/intechopen.76249
• Arias, C., Soliverez, C., & Bozzi, N. (2020). El envejecimiento poblacionalen América
Latina: Aportes para el delineamiento de polfticas publicas. Revista Euro latinoamericana de
Analisis Social y Polftico (RELASP), 1(2), 11-23. unr.edu.ar
• Arroyo, L. I., Ortega-Lenis, D., Ardila, L., Leal, F., Morales, O., Calvache, J. A., & De
Vries, E. (2022). Percepções dos médicos sobre os cuidados de fim de vida em pacientes
oncológicos. Revista Gerencia y Polfticas de Salud, 21, 1-21. redalyc.org.
• Athar, A., Dhaduk, K., & Aronow, W. (2019). Doenças do pericárdio em pacientes
idosos ... https://doi.org/10.5772/intechopen.89473
• Ballesteros, F. Z. (2023). OSTEOARTRITE DO JOELHO, SAÚDE OSTEOMUSCULAR E
ACTIVIDADE FÍSICA NOS IDOSOS. EmasF, Revista Digital de Educaci6n Ffsica,
15(85). webcindario.com
• BALTASAR, M. C., MENDOZA, M. D., & LOPEZ-REY, C. A. (). REVISÃO DA
LITERATURA: DISFAGIA NO ADULTO IDOSO.
publicacionescientificas.es. publicacionescientificas.es

• Bautista-Mier, H., Rodrfguez-Gutiérrez, A., Torres-Espinosa, C., & L6pez- Ramfrez, J.
(2021). Uso e perceção de um aplicativo móvel para avaliação geriátrica abrangente por
profissionais de saúde. Medunab, 24(2), 169-
182. https://doi.org/10.29375/01237047.4041

• Beracasa, L., Bar6n, C., & Sanchez, J. (2021). Toxicidade relacionada ao tratamento do
câncer em adultos mais velhos. revisão da literatura. Universitas Médica, 62(1).
https://doi.org/10.11144/javeriana.umed62- 1.toxi
• Bertolotti, L. (2022). Funcionamento cognitivo no envelhecimento: intervenção
psicopedagógica. ufasta.edu.ar
• Bora, A., Koç, M., Durmus, K., & Altunta□, E. (2021). Avaliação da frequência das
variações anatómicas da região nasossinusal na idade pediátrica e adulta.

grupos de acordo com o género: resultados de tomografia computorizada de 1532 casos. The Egyptian Journal of Otolaryngology, 37(1). https://doi.org/10.1186/s43163-021-00122-9

• Burgos, L. E. M., Âlvarez, R. E. Z., Bermudez, L. S. M., & Cedeño, C. I. M. (2023). Prevenção de doenças crónicas avançadas e saúde pública. RECIAMUC, 7(2), 55-64. reciamuc.com.

• Camacho Torres, L. V. & Trejo Bravo, F. V. (2023). La ansiedad en el proceso de envejecimiento de los/as adultos/as mayores del Centro De Salud Tipo C "Bastión Popular" de la ciudad de Guayaquil en el periodo de ... ups.edu.ec

• Castro Silva, J. Y. (2022). Propuesta de modelo de atención con enfoque de desarrollo humano sostenible para mejorar la calidad de vida de adultos mayores. Chachapoyas. untrm.edu.pe

• Cedillo, G. J. (2020). Trabalho social em saúde: teoria e práxis inovadora. Margen: revista de serviço social e ciências sociais. academia.edu

• CISNEROS, L. O. (2021). Atividade da telomerase e regulação do ciclo celular durante a transição da fase de neurogénese para a fase de gliogénese na medula espinhal. unam.mx

• Collazo, C., Santos, J. G., Bernal, J. G., & Cubo, E. (2020). Situação do estado da utilização e potenciais utilidades das novas tecnologias para medir a atividade física. Revisão sistemática da literatura. Atenção Práticas Primárias. sciencedirect.com

• Criollo, W. (2019). Avaliação da capacidade funcional e das atividades de vida diária em idosos institucionalizados. Movimento Científico, 13(2). https://doi.org/10.33881/2011-7191.mct.13201

• Duque-Fernandez, L. M., Ornelas-Contreras, M., & Benavides-Pando, E. V. (2020). Atividade física e sua relação com o envelhecimento e a capacidade funcional: uma revisão da literatura de pesquisa. Psicologfa y Salud, 30(1), 45-57. uv.mx

• Furuya, J., Tamada, Y., Sato, T., Hara, A., Nomura, T., Kobayashi, T., - & Kondo, H. (2015). O uso de próteses completas está associado a alterações na forma tridimensional da orofaringe em idosos edêntulos que afectam a deglutição. Gerodontology, 33(4), 513-521. https://doi.org/10.1111/ger.12197

• Gallegos, W., Toia, A., & Rivera, R. (2021). Analisis psicométrico de la escala de depresión geriatrica de yesavage en adultos mayores de la macroregión sur del peru. Revista Enfermeria Herediana, 12, 11-19. https://doi.org/10.20453/renh.v12i0.3960

• Garcfa Montes de Oca, A. L. & in Estomatologfa, E. P. G. (). A DETERIORAÇÃO DA FUNÇÃO MASTIGATÓRIA E A INFLUÊNCIA DA NUTRIÇÃO NO ADULTO IDOSO. A Deterioração da função mastigatória ... cisalud-. ucmh.sld.cu. sld.cu

• Gondim, E., Moreira, M., Lima, A., Aquino, P., & Nascimento, S. (2022). As mulheres sabem sobre o risco de trauma perineal, mas não sabem como preveni-lo: conhecimento, atitude e prática. International Journal of Gynecology & Obstetrics, 161(2), 470-477. https://doi.org/10.1002/ijgo.14526

• Gonzalez-Montalvo, J. I., Ramfrez-Martfn, R., Colino, R. M., Alarcón, T., Tarazona-

Santabalbina, F. J., Martfnez-Velilla, N., ... & Martfn-Sanchez, F. J. (2020). Geriatria transversal. Um desafio de saúde para o século XXI. Revista Espafiola de Geriatrfa y Gerontologfa, 55(2), 84-97. [HTML].

• Grande Aranda, J. I. (2024). Os vulneráveis: estudos interdisciplinares s o b r e vulnerabilidade. [HTML].

• Huerta Pozo, K. (2021). Influencia de cambios sociales en los sentimientos del proceso de envejecimiento "Centro Integral del Adulto Mayor"- Municipalidad Provincial de Huanuco 2019. 200.37.135.58

• Iglesias, S. R., Preciado, M. C. R., & Carrasco, L. Y. V. (2024). Impacto del Programa Social Pensi6n 65 en la satisfacci6n de necesidades de adultos mayores peruanos. Revista InveCom/ISSN en lfnea: 2739-0063, 4(2), 1-10. revistainvecom.org.

• Jernigan, T., Archibald, S., Berhow, M., Sowell, E., Foster, D., & Hesselink, J. (1991). Cerebral structure on mri, part i: localization of age-related changes. Biological Psychiatry, 29(1), 55-67. https://doi.org/10.1016/0006- 3223(91)90210-d

• Juna, H. e Paez, M. (2023). Criação e validação do formato de avaliação de saúde de enfermagem para pacientes geriátricos. Revista Brasileira de Revisão em Saúde, 6(1), 2815-2827. https://doi.org/10.34119/bjhrv6n1-221

• Kanasi, E., Ayilavarapu, S., & Jones, J. (2016). O envelhecimento da população: demografia e a biologia do envelhecimento. Periodontology 2000, 72(1), 13-18. https://doi.org/10.1111/prd.12126

• LA, V. EPÍLOGO FINAL. wordpress.com

• Landazabal, O. S. & Barboza, F. Y. A. (2020). Derechos humanos del adulto mayor en el ambito familiar colombiano en el marco del envejecimiento demografico. Jurfdicas CUC. unirioja.es

• Lemaître, H., Goldman, A., Sambataro, F., Verchinski, B., Meyer-Lindenberg,

A., Weinberger, D., - & Mattay, V. (2012). Cérebro normal relacionado à idade

alterações morfométricas: não uniformidade na espessura cortical, na área de superfície e no volume da substância cinzenta? Neurobiology of Aging, 33(3), 617.e1-617.e9. https://doi.org/10.1016/j.neurobiolaging.2010.07.013

• Malan Valente, A. B. (2024). Humanizaci6n en el cuidado del adulto mayor en el area de hospitalizaci6n del Hospital Basico Guamote. udla.edu.ec

• Martfnez, D., Martfnez, A., Sanchez, M., & Ramos, S. (2022). Importancia de la valoraci6n geriatrica integral a prop6sito de un caso. Investigaci6n Y Desarrollo, 10(1), 56-60. https://doi.org/10.31243/id.v10.2016.182

• Mazzini, M. B. B., Salazar, M. F. B., Sanchez, J. P. E., & Amaya, J. E. R. (2023). Avaliação geriátrica abrangente para adultos mais velhos. Polo del Conocimiento: Revista cientffico-profesional, 8(6), 1453-1473. unirioja.es

• Montesino, D. C., Reguera, I. P., Fernandez, O. R., Relova, M. R., & Valladares,

W. C. (2022). Caracterização clínica e epidemiológica da deficiência na população idosa. Interdisciplinar

Rehabilitation/Rehabilitacion Interdisciplinaria, 2, 15-15. saludcyt.ar

• Mufioz Toledo, M. B. & Ochoa Cueva, I. S. (2023). Calidad de vida del adulto mayor en una comunidad segun Betty Neuman. ucacue.edu.ec

• Mufioz Zuta, J. L. (). Conceção de um sistema de monitorização inteligente intradomiciliario y remoto para el cuidado de la población adultaci6n mayor. tesis.pucp.edu.pe. pucp.edu.pe

• Ney, D., Weiss, J., Kind, A., & Robbins, J. (2009). Deglutição na senescência: impacto, estratégias e intervenções. Nutrition in Clinical Practice, 24(3), 395-413. https://doi.org/10.1177/0884533609332005

• Nigam, Y., Knight, J., Bhattacharya, S., & Bayer, A. (2012). Alterações fisiológicas associadas ao envelhecimento e imobilidade. Journal of Aging Research, 2012, 1-2. https://doi.org/10.1155/2012/468469

• Nufiez, D. (2017). Avaliaçãc geriátrica abrangente em adultos mais velhos com cancro. Revista Clfnica Escuela De Medicina Ucr-HSJD, 7(3). https://doi.org/10.15517/rc_ucr-hsjd.v7i3.30018

• Ocampo, D., Mufioz, I., & G6mez, F. (2017). Previsão de medidas de desempenho em idosos institucionalizados em cadeira de rodas. Revista Colombiana De Rehabilitaci6n, 12(1), 6. https://doi.org/10.30788/revcolreh.v12.n1.2013.45

• Olivares Ramfrez, F. (2021). Perceção da funcionalidade familiar e apoio social em pacientes com deficiência devido à Covid 19 na Delegação HGZ1 Aguascalientes. uaa.mx

• Orozco, J. C., Diaz, J. R. D. L. O., & Brenes, A. G. M. (2024). Aproximaci6n a la Intervenci6n de la Demencia desde los Cuidados Paliativos. Ciencia Latina: Revista Multidisciplinar, 8(1), 690-703. unirioja.es

• Parrales, G. L. P. & Molina, S. A. A. (2020). A família no cuidado do idoso. utm.edu.ec

• Pinilla, J. M. G., Dfez-Villanueva, P., Freire, R. B., Formiga, F., Marcos, M. C., Bonanad, C., ... & Martfnez-Sellés, M. (2020). Documento de consenso e recomendações sobre cuidados paliativos na insuficiência cardíaca das Secções de Insuficiência Cardíaca e Cardiologia Geriátrica da Sociedade Espanhola de Cardiologia. Revista espafiola de cardiologfa, 73(1), 69-77. [HTML]

• Pifia Moran, M., Olivo Viana, M. G., Martfnez Matamala, C., Poblete Troncoso, M., & Guerra Guerrero, V. (2022). Envelhecimento, qualidade de vida e saúde. Desafios aos papéis sociais das pessoas idosas. Rumbos TS, 17(28), 7-27. scielo.cl

• Ramfrez, G. E. R. (2023). La Biologfa Molecular del envejecimiento. Ciencia Latina Revista Cientffica Multidisciplinar. ciencialatina.org

• Reyes Sanchez, L. P. & Sandoval Bocanegra, V. A. (2023). Stress, depressão e apoio social em adultos mais velhos. ucv.edu.pe

• Reyna, R., Contreras, M., & Vega, H. (2021). Uso do Questionário de Saúde SF-36 em adultos mais velhos. Revisão sistemática. Ansiedade e stress. researchgate.net

• Salas Zapata, C., Aguilar G6mez, M., Giraldo Sanchez, T. E., Mufioz Rua, M. C., Torres Bland6n, A., Uribe Castafio, A., & Uribe Quintero, A. (2021). Depressão maior na população geral de Envigado (Colômbia): prevalência e fatores associados. CES Psicologfa, 14(3), 117-133. scielo.org.cc

• Scilimati, N., Beccati, F., Dall'Aglio, C., Meo, A., & Pepe, M. (2022). A idade e o sexo estão correlacionados com alterações ósseas e variações anatómicas da região

lombossacroilíaca da coluna vertebral numa população mista de cavalos. Journal of the American Veterinary Medical Association, 1-8.
https://doi.org/10.2460/javma.22.07.0293. https://doi.org/10.2460/javma.22.07.0293

• Tomala, F. & Rivera, S. N. Y. (2023). Abandono familiar e estado emocional de adultos mais velhos no bairro de Parafso, no cantão de Salinas.... 593 Digital
Editora CEIT. unirioja.es

• Torres, A., Pérez-Galavfs, A., Ron, M., & Mendoza, N. (2023). Fatores psicossociais do local de trabalho e estresse no pessoal de assistência médica.
Interdisciplinary Rehabilitation/Rehabilitacion Interdisciplinaria, 3, 42-42.

healthcyt.co.uk

• Vega Baudrit, J., Corrales, R., Castillo Henrfquez, L., & Camacho, M. (2024). Telemedicina e a Internet das Coisas Médicas (IoMT): Superando desafios e visualizando oportunidades para melhorar a qualidade de vida e a acessibilidade no sector da saúde.
ulead.ac.cr

• Velasco Gaibor, C. E. (2022). Abandono da família e sua influência no estado emocional de uma mulher de 65 anos hospitalizada no Centro Residencial Gerontol6gico Atalaya no cantão de Chillanes.... 190.15.129.146

• Wanden-Berghe, C. (2021). Avaliação geriátrica global. Hospital em Casa, 5(2), 115.
https://doi.org/10.22585/hospdomic.v5i2.136

• Yungplut, F. (2024). Estratégias diagnósticas e terapêuticas utilizadas em pacientes não institucionalizados com mais de 70 anos de idade baseadas na perda funcional das capacidades físicas... ufasta.edu.ar

• Zarebski, G. (2021). A Organização Mundial de Saúde (OMS): Do Envelhecimento Saudável à Velhice como Doença. Desafios para a Gerontologia. Revista IGERMED.
inicien.com

• Zubiria, L. M. A. (2024). Desaffos y perspectivas de la polftica publica del envejecimiento en Colombia. Revista Venezolana de Gerencia: RVG. unirioja.es

CAPÍTULO 2
B-APRENDIZAGEM

Nas últimas décadas, foram empregues várias abordagens diferentes para utilizar as TIC (tecnologias da informação e da comunicação) para apoiar o ensino e a aprendizagem no ensino superior.

A tendência é para a integração da tecnologia móvel (tablets, smartphones, servidores portáteis) com abordagens pedagógicas flexíveis, tais como salas de aula integradas, espaços de criação construtivos e gamificação (Becker et al. 2018; Johnson et al. 2014).

As tendências da educação virtual registaram um desenvolvimento exponencial desde a pandemia de COVID 19, o último Horizon Report (Alexander et al. 2019) descreveu uma análise situacional a favor de tendências como a aprendizagem móvel e as tecnologias analíticas; e, num futuro próximo, espera-se que a inteligência artificial (IA) desempenhe um papel mais importante na educação. Esta situação já está a ocorrer hoje em dia, e prevê-se que os desenvolvimentos num futuro próximo se centrem na cadeia de blocos e nos assistentes virtuais para apoio à aprendizagem educativa.

É neste contexto que surge o E lerning (e-learning) que permite sessões síncronas e assíncronas e a repetição de conteúdos para uma melhor fixação visual (Tan e Erdogan, 2004; Yal n, 2000). No entanto, o e-learning assíncrono sofre de algumas limitações amplamente descritas na literatura, tais como a sensação de isolamento e a falta de motivação dos alunos (Dogan, Duman & Seferoglu, 2011), bem como a fraca comunicação e interação social com o professor e entre pares.

Assim, o B-learning surgiu como um processo de aprendizagem que integra aspectos da aprendizagem presencial e em linha através da utilização de tecnologias de e-learning aplicadas a ambientes de aprendizagem tradicionais.

Os seus benefícios foram rapidamente percebidos, tais como uma velocidade efectiva de aprendizagem individual e a criação de ambientes de aprendizagem mais flexíveis.

O B leraning pode ser definido como uma abordagem educativa que combina o ensino tradicional presencial com actividades de aprendizagem em linha (Amponsah, 2020). Caracteriza-se pela integração da tecnologia no processo de aprendizagem, permitindo uma experiência de aprendizagem flexível e personalizada (Valtonen et al., 2020).

A aprendizagem B fornece apoio pedagógico através de uma variedade de métodos, incluindo palestras, questionários formativos, avaliação automatizada, autoavaliação e avaliação pelos pares, e fóruns online para apoio e discussão entre pares (Glance et al, 2013). Esta combinação de estratégias de ensino promove a aprendizagem ativa, a participação e a colaboração entre os alunos (Boston-Hill et al., 2021).

Características

https://linnealab.wordpress.com

As principais características da aprendizagem mista são (Dangwal, K. L. (2017)...:

• Os alunos têm a oportunidade de utilizar ambas as modalidades: o ensino misto permite, na sua componente tradicional, uma interação pessoal com o professor e os colegas e complementa a sua aprendizagem com o apoio das TIC.

• Isto depende em grande parte da natureza do conteúdo e dos objectivos, que são concebidos pelos professores, que escolhem o método adequado.

• Para uma aplicação correcta, os professores devem estar familiarizados com ambas as metodologias. Uma caraterística fundamental é o dinamismo e a formação técnica para migrar entre o formato tradicional de sala de aula e o formato online.

o formato apoiado pelas TIC. Isto está intimamente relacionado com a disponibilidade de infra-estruturas e recursos de TI.

• Os estudantes desenvolvem competências na utilização de novas tecnologias e adquirem a capacidade de explorar ao máximo as tecnologias disponíveis.

• Permite o desenvolvimento integral da personalidade. Em domínios como o cognitivo, o físico e o emocional. O ensino tradicional em sala de aula é útil ao nível da memória e da compreensão cognitiva e é complementado por actividades em linha. As experiências autogeridas ajudam ao nível reflexivo da aprendizagem.

• Os alunos adquirem uma ampla exposição e novas perspectivas sobre o conteúdo da disciplina, o seu conhecimento do conteúdo é enriquecido, podem desenvolver novas dimensões de prática e de aprendizagem significativa.

• O papel do professor na aprendizagem mista é diversificado para além do tradicional; torna-se um motivador, facilitador, organizador e criador de conteúdos através das TIC.

Têm a oportunidade de desenvolver o seu crescimento profissional.

• O aluno constrói o conhecimento em vez de simplesmente o repetir; é desenvolvido um modelo de construtivismo em que os alunos se tornam autogeridos e auto-eficazes no seu desenvolvimento académico e pessoal.

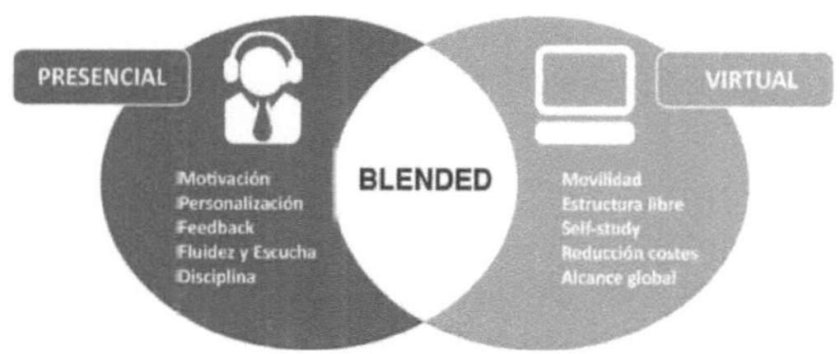

Benefícios

Uma das vantagens da aprendizagem B no ensino superior é a sua capacidade de atender a diferentes estilos e preferências de aprendizagem (Amponsah, 2020). Ao incorporar componentes em linha e presenciais, a aprendizagem B permite que os estudantes interajam com os materiais do curso de uma forma adaptada aos seus estilos de aprendizagem individuais (Amponsah, 2020). às suas necessidades individuais. Esta flexibilidade promove a aprendizagem autónoma e permite que os estudantes se apropriem da sua educação (Amponsah, 2020). Além disso, o B-learning oferece oportunidades para os estudantes desenvolverem competências de literacia digital, que são essenciais na atual sociedade orientada para a tecnologia (Vlachopoulos & Makri, 2017).

O estudo de Cobanoglu, A. et al (2014) demonstrou benefícios em 2 áreas principais: Desempenho académico, embora não significativamente em relação à atividade presencial (dependendo das concepções pedagógicas implementadas e dos factores motivacionais); Perceção da flexibilidade cognitiva, os alunos desenvolveram actividades metacognitivas e descobriram diferentes pontos de vista ao interagir com os pares.

Limitações

No entanto, a aprendizagem online também tem as suas limitações. Uma limitação é o potencial acesso desigual à tecnologia e à conetividade à Internet, o que pode criar disparidades nas oportunidades de aprendizagem (Meskhi et al., 2019). Outra limitação é a necessidade de uma gestão eficaz do tempo e de competências de autorregulação, uma vez que a aprendizagem B exige que os aprendentes assumam a responsabilidade pela sua própria aprendizagem e se mantenham motivados (Valtonen et al., 2020). Além disso, a conceção e a implementação de cursos B-learning requerem um planeamento cuidadoso e a consideração de estratégias de ensino para garantir que as componentes online e presenciais sejam perfeitamente integradas (Herrington e Herrington, 2006).

Perspectivas

Apesar destas limitações, a aprendizagem em linha tem perspectivas promissoras no ensino superior. Oferece o potencial de melhorar a qualidade da educação, combinando os melhores aspectos do ensino tradicional com as vantagens da aprendizagem em linha (Valtonen et al., 2020). O B-learning pode também facilitar a colaboração e a comunicação entre os estudantes e entre estes e os professores, promovendo um sentido de comunidade e de envolvimento (Boston-Hill et al., 2021). Além disso, o B-learning pode proporcionar oportunidades de aprendizagem ao longo da vida e de desenvolvimento profissional, uma vez que permite aos indivíduos aceder a recursos educativos e participar em cursos a partir de qualquer lugar e em qualquer altura (Amponsah, 2020).

Conclusão

Em conclusão, o B-learning no ensino superior é uma abordagem pedagógica que combina o ensino presencial com actividades de aprendizagem em linha. Fornece apoio pedagógico através de uma variedade de métodos e promove a aprendizagem ativa e a colaboração. A aprendizagem B oferece vantagens como a flexibilidade, a aprendizagem personalizada e o desenvolvimento de competências de literacia digital. No entanto, também tem limitações relacionadas com o acesso à tecnologia e a necessidade de uma gestão eficaz do tempo. Apesar destas limitações, a aprendizagem B tem perspectivas promissoras no ensino superior, incluindo a melhoria da qualidade da educação, a promoção da comunidade e da participação e a facilitação da aprendizagem ao longo da vida e do desenvolvimento profissional.

REFERÊNCIAS

• Amponsah, S. (2020). Explorando os estilos de aprendizagem dominantes dos alunos adultos no ensino superior. International Review of Education, 66(4), 531-550. https://doi.org/10.1007/s11159-020-09845-y

• Becker, S., Brown, M., Dahlstrom, E., Davis, A., DePaul, K., Diaz, V., et al. (2018). Relatório NMC Horizon: 2018 superior (educação ed.). Louisville, CO: EDUCAUSE.

• Boston-Hill, K., Stelljes, D., Boersma, J., & Boersma, J. (2021). Debate para a aprendizagem cívica: um modelo para renovar a missão cívica do ensino superior. Journal of the Scholarship of Teaching and Learning, 21(4). https://doi.org/10.14434/josotl.v21i4.32845

• Cobanoglu, A., & Yurdakul, B. (2014). The Effect of Blended Learning on Students' Achievement, Perceived Cognitive Flexibility Levels and Self- Regulated Learning Skills*. Revista de Educação e Prática, 5, 176-196.

• Dangwal, K. L. (2017). Aprendizagem combinada: Uma abordagem inovadora. Revista Universal de Investigação Educacional, 5(1), 129-136.

• Glance, D., Forsey, M., & Riley, M. (2013). Os fundamentos pedagógicos dos cursos online abertos e massivos. First Monday. https://doi.org/10.5210/fm.v18i5.4350

• Herrington, A. e Herrington, J. (2006). O que é um ambiente de aprendizagem autêntico?, 1-14. https://doi.org/10.4018/978-1-59140-594-8.ch001

• Meskhi, B., Ponomareva, S., & Ugnich, E. (2019). E-learning no ensino superior inclusivo: necessidades, oportunidades e limitações. Revista Internacional de Gestão Educacional, 33(3), 424-437.https://doi.org/10.1108/ijem-09-2018-0282

• Valtonen, T., Leppanen, U., Hyypia, M., Kokko, A., Manninen, J., Vartiainen, H.,& Hirsto, L. (2020). Ambientes de aprendizagem preferidos pelos estudantes universitários: uma mudança para ambientes de aprendizagem informais e flexíveis. LearningEnvironments Research, 24(3), 371-388. https://doi.org/10.1007/s10984-020-09339-6

• Vlachopoulos, D. e Makri, A. (2017). O efeito dos jogos e simulações no ensino superior: uma revisão sistemática da literatura. Revista Internacional de Tecnologia Educacional no Ensino Superior, 14(1). https://doi.org/10.1186/s41239-017-0062-1

ENSINO DE SÍNDROMES GERIÁTRICAS A MÉDICOS RESIDENTES ATRAVÉS DE B-LEARNING

Introdução

No domínio da medicina geriátrica, a formação adequada na gestão das síndromes geriátricas é crucial devido ao número crescente de adultos mais velhos e às complexas necessidades de saúde associadas a esta população. As síndromes geriátricas, como a imobilidade, a sonolência, a incontinência urinária, o défice cognitivo, a sarcopenia e a fragilidade, apresentam desafios únicos que exigem uma abordagem especializada e abrangente. Este livro foi concebido para dotar os médicos residentes dos conhecimentos e competências necessários para diagnosticar, tratar e gerir eficazmente estas síndromes, utilizando o método B-learning.

O B-learning ou aprendizagem mista combina elementos da aprendizagem tradicional presencial e técnicas de e-learning para criar uma experiência educativa mais rica e flexível. Esta abordagem facilita uma compreensão mais profunda e uma aprendizagem autónoma, permitindo que os residentes interajam tanto com os conteúdos digitais como com a experiência prática. Neste contexto, o B-learning é particularmente adequado ao ensino das síndromes geriátricas, uma vez que permite a incorporação de múltiplos recursos educativos, tais como simulações, videoconferências, fóruns de discussão e actividades práticas. O ensino das síndromes geriátricas aos médicos residentes é um aspeto fundamental da sua educação e formação. As síndromes geriátricas, que ocorrem frequentemente na população geriátrica, são condições clínicas que afectam a saúde e o bem-estar dos adultos mais velhos. O objetivo deste curso é fornecer aos médicos residentes os conhecimentos necessários para reconhecer, diagnosticar e tratar adequadamente as diferentes síndromes geriátricas. Através da abordagem b-learning, que combina o ensino presencial com o e-learning, o curso foi concebido para fornecer os conhecimentos necessários para reconhecer, diagnosticar e tratar adequadamente as diferentes síndromes geriátricas. O objetivo é proporcionar uma formação abrangente e actualizada que permita aos médicos residentes adquirir as competências necessárias para um tratamento ótimo dos doentes geriátricos. Esta secção introduzirá o tema, destacando a importância do ensino das síndromes geriátricas e a sua relação com o b-learning.

Este capítulo abordará cada síndrome geriátrica específica e foi concebido para servir como um guia abrangente para um ensino e aprendizagem eficazes. No final de cada capítulo, os médicos residentes não só terão adquirido conhecimentos teóricos, como também terão desenvolvido competências práticas através de exercícios interactivos e estudos de casos concebidos para reproduzir desafios do mundo real. Além disso, a integração das tecnologias de informação e das plataformas de e-learning garante que os médicos residentes possam continuar a sua formação de forma contínua e adaptável, respondendo às novas exigências do campo da geriatria. Este capítulo tem como objetivo não só educar, mas também inspirar os futuros geriatras a adoptarem uma abordagem holística e empática aos cuidados dos idosos, melhorando assim a sua qualidade de vida e a sua dignidade. Com ênfase na inovação educacional e na aplicação prática, pretendemos estabelecer um novo padrão na formação médica geriátrica.

1. Definição de Síndromes Geriátricas

As síndromes geriátricas referem-se a um conjunto de doenças multifactoriais e comuns na população geriátrica que se caracterizam pela presença de sinais e sintomas clínicos específicos. São condições heterogéneas que incluem a imobilidade, os cafés, o défice cognitivo, a sarcopenia e a fragilidade.Estas síndromes ocorrem como resultado da interação de múltiplos factores e o seu diagnóstico baseia-se na identificação dos critérios clínicos correspondentes.É importante salientar que estas síndromes não devem ser consideradas como doenças isoladas, mas sim como manifestações clínicas múltiplas e inter-relacionadas. A sua abordagem e tratamento requerem uma abordagem multidisciplinar que envolva diferentes especialidades médicas, a fim de melhorar a qualidade de vida dos doentes geriátricos e prevenir complicações futuras. Neste sentido, a colaboração entre geriatras, fisioterapeutas, nutricionistas, psicólogos e assistentes sociais é crucial para estabelecer planos de cuidados abrangentes que abordem eficazmente cada síndrome individualmente e em conjunto. É essencial ter em conta que muitas destas síndromes geriátricas aumentam o risco de incapacidade, hospitalização e mortalidade nas pessoas idosas, pelo que a sua deteção precoce e gestão adequada são aspectos essenciais dos cuidados de saúde geriátricos. Por conseguinte, é crucial promover a sensibilização e a formação na gestão destas síndromes, tanto em contextos clínicos como comunitários, a fim de garantir cuidados abrangentes e de qualidade para a população geriátrica. (Pefia et al.2020)(Corzo Camacho).

2. Classificação das síndromes geriátricas

A classificação das síndromes geriátricas permite organizar e categorizar os diferentes tipos de condições médicas que afectam as pessoas idosas. Entre as síndromes geriátricas mais comuns, contam-se: a síndrome de imobilidade, caracterizada pela perda de movimento e de capacidade funcional; a síndrome de caudaemia, que se refere à caudaemia recorrente e às suas consequências nos idosos; a síndrome de deficiência cognitiva, que envolve a perda progressiva de capacidades mentais; a sarcopenia, que é a perda de massa muscular e de força nos idosos; e a síndrome de fragilidade, que se caracteriza por uma diminuição da resistência e da vulnerabilidade física. Cada uma destas síndromes tem a sua própria definição específica, epidemiologia, diagnóstico, instrumentos de avaliação e abordagem e tratamento específicos, que são essenciais para prestar cuidados geriátricos abrangentes e eficazes.

a. Síndrome de Imobilidade

• Definição e relevância clínica

A imobilidade na população idosa é definida como a perda parcial ou total da capacidade de se mover e realizar actividades físicas de forma independente. Esta síndrome geriátrica é de particular interesse clínico devido às suas profundas implicações na qualidade de vida e na saúde geral dos idosos. A imobilidade aumenta significativamente o risco de complicações como úlceras de pressão, trombose venosa, infecções respiratórias e deterioração psicossocial, entre outras. Caracteriza-se pela ausência de atividade física ou pela restrição de movimentos devido a várias condições subjacentes, como a doença crónica, a deficiência ou o défice cognitivo. De acordo com a epidemiologia, esta síndrome afecta uma elevada percentagem de pessoas idosas, especialmente em ambientes de cuidados de longa duração. O diagnóstico da síndrome de imobilidade baseia-se na avaliação clínica da mobilidade, da força muscular e da capacidade funcional. Para o efeito, são utilizados instrumentos de avaliação como as escalas de avaliação da mobilidade e testes físicos específicos. A abordagem e o tratamento da síndrome de imobilidade incluem intervenções não farmacológicas, como a fisioterapia, o exercício físico adaptado e a terapia ocupacional, que têm como objetivo melhorar a mobilidade e prevenir as complicações associadas à imobilidade. Além disso, podem ser consideradas estratégias farmacológicas para a gestão dos sintomas e controlo da dor.

É essencial que os médicos residentes adquiram conhecimentos sólidos sobre esta síndrome, uma vez que a sua deteção precoce e tratamento adequado são cruciais para melhorar a qualidade de vida e prevenir a incapacidade na população geriátrica (Tamayo Pérez, 2023) (Suarez Tomala, 2022).

• Factores de risco e consequências

Os factores de risco para a imobilidade incluem a idade avançada, doenças crónicas como a artrite e a doença de Parkinson, acidentes, operações cirúrgicas e estados pós-acidente vascular cerebral. A identificação precoce destes factores em doentes geriátricos é crucial para uma prevenção e gestão eficazes. As consequências da imobilidade não afectam apenas a saúde física dos idosos, mas também o seu bem-estar emocional e social, levando frequentemente a um ciclo de deterioração que pode ser difícil de inverter.

• Epidemiologia

A síndrome de imobilidade é um problema de saúde comum na população geriátrica. De acordo com estudos epidemiológicos, estima-se que cerca de 60% dos idosos institucionalizados e 30% dos idosos que vivem na comunidade sofram de síndrome de imobilidade. A prevalência da imobilidade aumenta com a idade, sendo mais comum em pessoas com mais de 80 anos de idade. Esta condição está associada a um risco acrescido de complicações médicas e a uma redução da qualidade de vida. A epidemiologia da síndrome de imobilidade sublinha a importância do seu reconhecimento e tratamento adequado na formação dos médicos residentes, com o objetivo de melhorar os cuidados prestados aos doentes geriátricos (Tamayo Pérez, 2023) (MUNOZ).

• Diagnóstico

O diagnóstico de imobilidade baseia-se na avaliação funcional do doente, utilizando instrumentos como a Escala de Mobilidade Timed Up and Go (TUG) e o Teste de Pé e

Marcha. Além disso, podem ser utilizados instrumentos de avaliação específicos, como o Teste de Tinetti, para avaliar a funcionalidade e o risco de quedas. Para além disso, é essencial uma avaliação abrangente que inclua aspectos cognitivos, emocionais e sociais para uma abordagem mais completa. O diagnóstico da Síndrome de Imobilidade baseia-se na avaliação clínica e na deteção de sinais e sintomas característicos da falta de mobilidade nos idosos. A história clínica do doente deve ser considerada, bem como uma avaliação física completa, incluindo testes de força muscular, equilíbrio e capacidade de realizar actividades básicas da vida diária. O diagnóstico precoce e preciso da Síndrome de Imobilidade é essencial para a implementação de uma abordagem e tratamento adequados, com o objetivo de melhorar a qualidade de vida e prevenir complicações em doentes geriátricos.

• Instrumentos de avaliação

Alguns dos instrumentos mais utilizados são a escala de Tinetti para avaliar o equilíbrio e a marcha, o teste Timed Up and Go para medir a mobilidade e o Mini-Mental State Examination (MMSE) para avaliar a função cognitiva. Estes instrumentos são de grande utilidade para o médico residente, uma vez que lhe permitem obter informações precisas e objectivas para o diagnóstico e acompanhamento das síndromes geriátricas, facilitando assim a implementação de uma abordagem adequada e de um tratamento eficaz.

• Abordagem e tratamento

A gestão da imobilidade requer uma abordagem multidisciplinar, incluindo fisioterapia para melhorar a força e o equilíbrio, intervenção nutricional para otimizar a saúde óssea e muscular e assistência psicológica para abordar as dimensões emocional e cognitiva. A implementação de tecnologias de assistência, como andarilhos e cadeiras de rodas, também desempenha um papel crucial na melhoria da autonomia do doente. (A abordagem e o tratamento da Síndrome de Imobilidade em doentes geriátricos requerem uma abordagem multidisciplinar. Devem ser implementadas intervenções farmacológicas e não farmacológicas com o objetivo de melhorar a mobilidade e prevenir complicações como as úlceras de pressão, o défice cognitivo e a diminuição da função cardiovascular e respiratória. As intervenções não farmacológicas incluem exercícios de mobilidade, fisioterapia, terapia ocupacional e programas de atividade física adaptada. Em termos de intervenções farmacológicas, podem ser utilizados medicamentos para aliviar a dor e a inflamação, bem como suplementos nutricionais para melhorar a saúde geral. É fundamental estabelecer um plano de cuidados individualizado que tenha em conta as necessidades e características de cada doente, promovendo a sua autonomia e qualidade de vida. Neste sentido, é essencial o trabalho em equipa entre médicos, enfermeiros, fisioterapeutas, terapeutas ocupacionais e outros profissionais de saúde. A educação do paciente e da família é também um elemento crucial na gestão da Síndrome de Imobilidade, pois permite-lhes compreender a importância de seguir as recomendações e participar ativamente nos cuidados do paciente. (Sanchez, 2023).

b. Síndrome de queda

A Síndrome de Cafdas é uma condição comum e grave na população geriátrica. É definida como a ocorrência de um ataque não intencional de cafeína que resulta em danos físicos, psicológicos ou sociais. É importante notar que a cauda equina pode ser considerada como um sintoma de um problema médico subjacente ou uma consequência das alterações associadas ao envelhecimento. A epidemiologia da síndrome das caudas mostra uma elevada prevalência em adultos mais velhos e representa uma causa significativa de morbilidade e mortalidade neste grupo etário. O diagnóstico da síndrome de caidas envolve a avaliação dos factores de risco e a identificação de eventos precipitantes. Instrumentos de avaliação como o Timed Up and Go (TUG) e o Teste de Berg são úteis para avaliar o equilíbrio e a marcha, bem como outros aspectos físicos relacionados com a coabitação. A abordagem e o tratamento da síndrome da cauda equina incluem estratégias preventivas e de reabilitação, que se baseiam na identificação e gestão dos factores de risco específicos presentes em cada doente. As intervenções não farmacológicas incluem terapias de exercício físico, fortalecimento muscular e programas de treino de equilíbrio (Cedefio et al.2024) (Franco & Lisbeth, 2022) É crucial que os médicos residentes compreendam a síndrome da cauda equina e adquiram as competências necessárias para o seu diagnóstico, gestão e prevenção, uma vez que isso contribuirá para melhorar a qualidade de vida e a segurança dos adultos mais velhos.

• Definição

As quedas na população idosa são definidas como eventos involuntários que fazem com que a pessoa acabe no chão ou noutro nível inferior, sem ser consequência de uma doença grave ou de um empurrão violento. Em geriatria, as quedas são consideradas indicadores-chave de fragilidade e de aumento do risco de morbilidades graves, bem como de diminuição da qualidade de vida.

É considerada uma condição muito comum na população geriátrica e representa um importante problema de saúde pública devido às consequências negativas que pode ter, como fracturas, incapacidade funcional e diminuição da qualidade de vida.

As Cafdas podem ser causadas por múltiplos factores, incluindo distúrbios da marcha, fraqueza muscular, perturbações do equilíbrio e condições médicas subjacentes.

Uma avaliação exaustiva é essencial para identificar as potenciais causas da fibrose quística e para estabelecer um plano de tratamento individualizado que inclua intervenções farmacológicas e não farmacológicas para prevenir a recorrência da fibrose quística e melhorar a segurança e a funcionalidade do doente (Cedefio et al.2024)(Tamayo Pérez, 2023).

• Epidemiologia e factores precipitantes

A Síndrome de Cafdas é uma condição altamente prevalente em adultos mais velhos, sendo considerada uma das principais causas de incapacidade e mortalidade neste grupo populacional. De acordo com estudos epidemiológicos, aproximadamente 30% dos adultos com mais de 65 anos de idade sofrem pelo menos um ataque de cafeína por ano.

Além disso, estima-se que o café seja responsável por mais de 50% das lesões acidentais em pessoas com mais de 75 anos de idade. As consequências do café podem ser graves, incluindo fracturas da anca, lesões na cabeça e diminuição da qualidade de vida. É importante referir que esta síndrome tem um impacto significativo nos sistemas de saúde, gerando custos elevados em termos de hospitalizações e cuidados médicos. Por conseguinte, são essenciais estratégias de prevenção e gestão adequadas para reduzir a incidência de cafeína nesta população vulnerável (Hart et al.2020)(Ganz & Latham, 2020).

Os factores precipitantes podem ser classificados como intrínsecos e extrínsecos. Os factores intrínsecos incluem problemas de saúde, tais como fraqueza muscular, distúrbios da marcha e do equilíbrio, deficiência cognitiva e efeitos secundários da medicação. Os factores extrínsecos envolvem condições ambientais, como iluminação inadequada, obstáculos na estrada e calçado inapropriado.

• Diagnóstico e avaliação do dcente idoso em risco de coabitação

O diagnóstico da Síndrome de Queda em adultos mais velhos inclui uma avaliação completa para identificar as causas subjacentes e os factores de risco. Isto envolve a revisão da história clínica do doente, a realização de testes físicos e funcionais, bem como a análise das condições ambientais em que o indivíduo se encontra. A avaliação de um doente idoso com risco de cauda equina deve ser holística e incluir uma revisão completa da história clínica, exame físico e testes de mobilidade. Instrumentos como a Escala de Equilíbrio de Berg e o Teste de Marcha de Seis Minutos são frequentemente utilizados para avaliar o risco de cauda equina. Estes instrumentos podem fornecer informações objectivas sobre o risco de cauda equina e podem detetar défices no equilíbrio e na marcha.

Um diagnóstico preciso ajudará a conceber um plano de tratamento personalizado que inclua intervenções destinadas a atenuar os factores de risco identificados e a melhorar a segurança e a mobilidade do doente (Iglesias et al.2022). Deve ser prestada especial atenção à revisão da medicação e à avaliação cognitiva, pois ambas podem contribuir significativamente para o risco de cofres. (COLCHADO ROSALES, 2021).

• Intervenções preventivas e tratamentos

A abordagem e o tratamento da Síndrome de Cafdas baseiam-se em várias estratégias. Inicialmente, devem ser identificados cs factores de risco modificáveis através de uma avaliação abrangente do doente, incluindo história clínica, medicamentos e outros factores de risco. O uso de medicamentos, distúrbios sensoriais e músculo-esqueléticos, entre outros, deve ser incentivado. O exercício físico regular deve ser encorajado, uma dieta equilibrada deve ser incentivada e os músculos devem ser fortalecidos através de programas de treino específicos.

Em termos de técnicas de prevenção de quedas, devem ser implementadas medidas como a remoção de obstáculos no ambiente, o uso de calçado adequado, a utilização de auxiliares de mobilidade e a promoção da segurança em casa.

Por fim, é essencial educar os doentes e os seus cuidadores sobre estratégias de prevenção de quedas e como atuar no caso de estas ocorrerem. É importante salientar que o tratamento da Síndrome de Cafdas deve ser personalizado, tendo em conta as necessidades e características de cada doente, com o objetivo de reduzir o risco de Cafdas e melhorar a qualidade de vida.As intervenções para prevenir a dependência de cafeína nos idosos devem ser multifactoriais e incluir ajustes no ambiente doméstico para eliminar riscos, programas de exercício para melhorar a força e o equilíbrio e rastreio farmacológico para minimizar os efeitos secundários que podem aumentar o risco de dependência de cafeína. Além disso, a educação sobre a prevenção da cafeína é crucial para os doentes e os prestadores de cuidados. Em alguns casos, podem ser recomendados dispositivos de assistência, como bengalas ou andarilhos, para melhorar a estabilidade. (COLCHADO ROSALES, 2021).

c. Incontinência urinária

A síndrome de incontinência nos idosos é definida como a perda involuntária de urina que pode afetar a qualidade de vida e a autonomia dos doentes.

A epidemiologia revela uma elevada prevalência, especialmente em mulheres idosas. O diagnóstico envolve a avaliação da história clínica, exames físicos, testes laboratoriais e questionários específicos para avaliar a incontinência. Os instrumentos de avaliação incluem o pad test, a escala internacional de gravidade da incontinência, entre outros. A abordagem e o tratamento da incontinência no idoso é multidisciplinar e inclui mudanças no estilo de vida, exercícios para o pavimento pélvico, dispositivos de contenção, medicamentos e, em alguns casos, procedimentos cirúrgicos.

• Definição

A incontinência urinária é definida como a perda involuntária de urina, um problema comum que afecta significativamente a dignidade e a qualidade de vida da população idosa. Esta condição não envolve apenas sofrimento clínico, mas também social e psicológico, com impacto na autonomia e na participação nas actividades diárias (Batmani et al., 2021) (Yagmur & Gül, 2021).

• Epidemiologia da síndrome de incontinência nos idosos

A incontinência afecta uma grande parte da população geriátrica. De acordo com estudos, cerca de 50% das pessoas com mais de 65 anos têm algum grau de incontinência. Para além disso, estima-se que 80% dos casos de incontinência urinária ocorram em mulheres. Esta informação é crucial para a formação dos médicos residentes, uma vez que lhes permite compreender a magnitude do problema e a necessidade de adquirir competências para um diagnóstico e tratamento eficazes na sua prática clínica. (Gibson et al.2021).

• Tipos de incontinência nos idosos

Nos idosos, a incontinência urinária pode ser classificada em vários tipos principais:

- Incontinência de esforço: perda de urina durante a tosse, espirros ou qualquer atividade que aumente a pressão intra-abdominal.
- Incontinência de urgência: perdas de urina associadas a uma forte vontade de urinar que é difícil de parar.
- Incontinência mista: uma combinação de incontinência de esforço e de urgência.
- Incontinência por extravasamento: ocorre quando a bexiga não se esvazia completamente, levando a episódios frequentes de micção ou a gotejamento contínuo.

• Diagnóstico diferencial e avaliação

O diagnóstico da Síndrome de Incontinência do Adulto Idoso é crucial para o tratamento adequado. É necessário utilizar instrumentos específicos para avaliar a incontinência urinária, tais como questionários de sintomas, diários miccionais, testes de absorventes, fluxometria, ecografia abdominal, urodinâmica, entre outros. É também fundamental a realização de uma história clínica detalhada, com especial atenção aos antecedentes médicos e cirúrgicos e à medicação atual. O exame físico deve incluir a avaliação genital, urinária e neurológica. Exames adicionais como urinálise, cultura microbiológica, ecografia renal, entre outros, devem ser considerados para excluir outras patologias concomitantes. A avaliação urodinâmica pode ser necessária para diferenciar os tipos de incontinência e para excluir outras condições médicas subjacentes. É essencial considerar factores como os medicamentos, as comorbilidades e o défice cognitivo, que podem influenciar a apresentação e a gestão da incontinência (Shaw & Wagg, 2021). Um diagnóstico atempado permitirá estabelecer um plano de gestão abrangente e personalizado para cada doente, tendo em conta factores como a causa subjacente, a gravidade, o impacto na qualidade de vida e os objectivos terapêuticos". (Radoja e Degmecié2020) (Nambiar et al.2022)

• Instrumentos de avaliação

No contexto da incontinência do adulto mais velho, os instrumentos de avaliação são uma parte crucial do processo de diagnóstico. Alguns dos instrumentos mais frequentemente utilizados incluem o Questionário de Incontinência Urinária (ICIQ), o Questionário de Avaliação da Incontinência (IAQ) e a Escala de Gravidade da Incontinência (IS) (Radoja e Degmecié.2020). Estes instrumentos permitem aos médicos residentes recolher informações detalhadas sobre a frequência, a gravidade e o impacto da incontinência na vida quotidiana dos doentes. Além disso, facilitam a identificação de possíveis causas subjacentes e contribuem para a formulação de um plano de tratamento adequado e personalizado para cada doente, o que é essencial para a gestão abrangente desta síndrome geriátrica" (O'Connor et al.2021). (O'Connor et al.2021)

• Opções de tratamento e gestão multidisciplinar

O tratamento da incontinência urinária nos idosos deve ser individualizado e pode incluir

- Modificações comportamentais: como o treino da bexiga e exercícios de fortalecimento do pavimento pélvico.
- Farmacoterapia: medicamentos específicos de acordo com o tipo de incontinência.
- Intervenções minimamente invasivas: tais como injecções de toxina botulínica ou dispositivos intravaginais.
- Cirurgia: em casos seleccionados em que outras terapias não tenham sido eficazes.

Esta gestão requer a colaboração de uma equipa multidisciplinar que inclua urologistas, geriatras, enfermeiros especializados, fisioterapeutas e psicólogos, assegurando cuidados abrangentes e centrados no doente.

d. Deficiência cognitiva

A Síndrome do Défice Cognitivo caracteriza-se por um declínio das capacidades cognitivas, como a memória, a linguagem, a atenção e o raciocínio, que afecta a capacidade da pessoa para realizar as actividades diárias. Esta síndrome é mais comum nas pessoas idosas e pode ser causada por diferentes doenças, como a doença de Alzheimer ou a demência vascular. A prevalência da síndrome do défice cognitivo aumenta com a idade e estima-se que afecte cerca de 20% dos adultos com mais de 65 anos de idade. O diagnóstico da síndrome baseia-se na avaliação clínica, na história do doente e em testes neuropsicológicos. Alguns dos instrumentos de avaliação utilizados incluem o Mini Exame do Estado Mental (MMSE) e a Avaliação Cognitiva de Montreal (MoCA). A gestão e o tratamento desta síndrome incluem intervenções farmacológicas e não farmacológicas, tais como medicamentos para melhorar a função cognitiva, terapia ocupacional e programas de estimulação cognitiva.

É essencial que os médicos residentes adquiram os conhecimentos necessários sobre esta síndrome, a fim de efectuarem um diagnóstico preciso e proporcionarem um tratamento adequado aos seus doentes.

• Definição e epidemiologia

A Síndrome de Deficiência Cognitiva é definida como um conjunto de sintomas cognitivos e funcionais que afectam a memória, a linguagem, a atenção, o raciocínio e a capacidade de realizar actividades da vida diária.

Estes sintomas têm um início gradual e progressivo e muitas vezes interferem significativamente com a independência e a qualidade de vida das pessoas idosas. Esta síndrome não é uma doença em si, mas pode ser causada por uma variedade de condições, como a doença de Alzheimer, a demência vascular e outras doenças neurodegenerativas. Uma avaliação exaustiva que inclua testes neuropsicológicos e escalas de classificação é essencial para estabelecer um diagnóstico preciso... (Chorefio-Parra et al.2020)(Parada et al.2022). A epidemiologia da Síndrome do Comprometimento Cognitivo é de grande relevância para compreender a prevalência e o impacto desta condição na população geriátrica. De acordo com estudos epidemiológicos, estima-se que cerca de 10-20% dos adultos mais velhos sofram de algum grau de comprometimento cognitivo, o que constitui um fator de risco para a demência. Além disso, verificou-se que a idade avançada e a presença de certas doenças crónicas, como a hipertensão e a diabetes, aumentam a probabilidade de desenvolver demência. Esses dados ressaltam a importância de se abordar adequadamente o diagnóstico e o tratamento da Síndrome do Comprometimento Cognitivo nos médicos residentes, a fim de melhorar a qualidade de vida dos pacientes geriátricos e reduzir os encargos que essa condição impõe aos sistemas de saúde. (Fonte Sevillano & Santos Hedman, 2020) (Parada et al., 2022).

À medida que a população mundial envelhece, o impacto da deficiência cognitiva continua a aumentar, representando um desafio significativo tanto para os sistemas de saúde como para as famílias das pessoas afectadas.

• Classificação e etiologia do défice cognitivo

O défice cognitivo pode ser classificado em duas categorias principais: défice cognitivo ligeiro (ICM) e demência. O DCL caracteriza-se por um declínio cognitivo que não interfere significativamente com as actividades diárias, enquanto a demência envolve uma perda da função cognitiva suficiente para interferir com a independência do indivíduo. As causas do declínio cognitivo são variadas, incluindo factores genéticos, doenças neurológicas, como a doença de Alzheimer e a demência vascular, e factores de risco modificáveis, como a hipertensão, a diabetes, o colesterol elevado, a falta de exercício físico e uma alimentação deficiente. Além disso, factores psicossociais como o isolamento social e a depressão podem também contribuir para o desenvolvimento do declínio cognitivo.

• Ferramentas de avaliação cognitiva

O diagnóstico da Síndrome de Deficiência Cognitiva é efectuado através de uma avaliação exaustiva que inclui testes neuropsicológicos, avaliação clínica e entrevistas com o doente e os seus cuidadores. São utilizados instrumentos de avaliação específicos. como o Mini-Exame do Estado Mental (MMSE) e a Classificação Clínica de Demência (CDR), para medir o grau de comprometimento cognitivo. Para além disso, podem ser realizados exames de neuroimagem, como a ressonância magnética cerebral, para excluir outras causas de défice cognitivo. É importante ter em conta que o diagnóstico deve ser efectuado por um médico com formação adequada, com base numa avaliação estruturada. uma vez que a Síndrome de Comprometimento Cognitivo pode estar associada a diferentes tipos de demência, como a doença de Alzheimer. (Cancino et al.2020).

Um diagnóstico precoce e exato permitirá estabelecer estratégias de gestão e tratamento adequadas para melhorar a qualidade de vida dos doentes e dos seus prestadores de cuidados (Galvez, 2024) (Parada et al.2022).

São utilizados vários instrumentos e testes neuropsicológicos para diagnosticar e avaliar o défice cognitivo. Estes incluem:

- Mini-Mental State Examination (MMSE)**: teste breve utilizado para avaliar a função cognitiva.
- Avaliação Cognitiva de Montreal (MoCA)**: concebida para identificar o ICM e a demência ligeira.
- Teste do Relógio: um teste simples e eficaz para avaliar a função executiva e visuoespacial.
- Além disso, podem ser utilizados questionários como o Informant Questionnaire on Cognitive Decline in the Elderly (IQCODE) para obter informações adicionais dos informadores dos doentes.

Estes instrumentos permitem uma avaliação objetiva e quantitativa do défice cognitivo, o que facilita um diagnóstico preciso e a monitorização do tratamento. É importante que os médicos residentes se familiarizem com estes instrumentos e aprendam a utilizá-los corretamente, a fim de realizarem uma avaliação completa e precisa dos pacientes com Síndrome de Comprometimento Cognitivo (Valenzuela Sanchez & Marcelino Arias, 2023) (Quinaloa et al.2020).A escolha do instrumento adequado depende da situação clínica e do grau de comprometimento suspeito. Estas avaliações devem ser complementadas com análises clínicas e, em alguns casos, com exames de imagem cerebral para excluir outras causas dos sintomas.

• Estratégias de cuidados e gestão dos doentes

A abordagem e o tratamento da Síndrome do Défice Cognitivo baseiam-se numa avaliação exaustiva das funções cognitivas e do estado geral de saúde do doente geriátrico. (Garcfa-Ribas et al.2023). O principal objetivo é retardar o declínio cognitivo e melhorar a qualidade de vida do doente. A abordagem e o tratamento da Síndrome de Comprometimento Cognitivo envolvem estratégias farmacológicas e não farmacológicas, como a estimulação cognitiva, a terapia ocupacional e o apoio emocional, com o objetivo de melhorar a qualidade de vida dos doentes e retardar a progressão dos sintomas. A utilização de fármacos anticolinesterásicos para melhorar os sintomas cognitivos deve ser considerada.

A gestão do défice cognitivo envolve uma abordagem multidisciplinar que inclui médicos, enfermeiros, terapeutas e assistentes sociais. As intervenções podem incluir:

- Gestão médica: ajuste de medicamentos que podem exacerbar o défice cognitivo e utilização de medicamentos para tratar sintomas específicos ou para retardar a progressão da demência.
- Intervenções psicossociais: tais como terapias de estimulação cognitiva, actividades de grupo e apoio emocional tanto para o doente como para os prestadores de cuidados.
- Modificações ambientais: garantir que o ambiente de vida do doente é seguro e

estimulante.

O trabalho em equipa com outros profissionais de saúde, como psicólogos e terapeutas, é essencial para prestar cuidados abrangentes adaptados às necessidades de cada doente. Também é necessário fornecer informação e apoio aos cuidadores e familiares, uma vez que estes desempenham um papel crucial na gestão e nos cuidados diários do doente com Síndrome de Deficiência Cognitiva (Sailema Lalaleo, 2023) (Mazo Bafiol & Moncada Botero, 2023).

e. Sarcopénia

A sarcopenia é uma síndrome geriátrica caracterizada pela perda progressiva e generalizada da massa muscular, da força e da função, e está associada a um risco acrescido de incapacidade, invalidez e diminuição da qualidade de vida na população idosa.

Estima-se que afecte aproximadamente 10% dos adultos com mais de 60 anos de idade e a sua prevalência aumenta com a idade. O diagnóstico baseia-se na avaliação clínica da força e da função muscular, bem como em análises laboratoriais para excluir outras causas de fraqueza. A abordagem e o tratamento da sarcopenia centram-se em intervenções multidisciplinares que incluem exercício físico, terapia nutricional e gestão dos factores de risco. É essencial que os médicos residentes adquiram conhecimentos sobre a sarcopénia e a sua gestão, de modo a identificar e gerir adequadamente esta condição na população geriátrica. (Pérez Carvajal, 2023) (Sandoval Animas).

• Definição e mecanismos fisicpatológicos

A sarcopenia caracteriza-se por um declínio progressivo e generalizado da massa muscular esquelética e da função muscular, que está associado a um risco acrescido de resultados adversos, como incapacidade, fracturas, dependência física e mortalidade. De um ponto de vista fisiopatológico, a sarcopenia resulta de um desequilíbrio entre a síntese e a degradação das proteínas musculares, um processo influenciado por factores como o envelhecimento, a inatividade física, a inflamação sistémica, a desnutrição e a desregulação endócrina.

• Epidemiologia

É uma das áreas de estudo mais importantes em relação às síndromes geriátricas. Estima-se que a prevalência da sarcopénia em adultos com mais de 60 anos varie entre 5% e 25%, dependendo dos critérios de diagnóstico utilizados. Além disso, observa-se que a sarcopenia tende a aumentar com a idade, sendo mais comum em pessoas com mais de 80 anos. Esta condição está associada a uma série de factores de risco, como a falta de atividade física, a desnutrição, a inflamação e doenças crónicas como a diabetes e a doença pulmonar obstrutiva crónica (Petermann-Rocha et al.2022)(Gao et al., 2021)(Almohaisen et al.2022).

• Diagnóstico de acordo com EWGSOP2

A mais recente atualização do Grupo de Trabalho Europeu sobre Sarcopenia em Pessoas Idosas (EWGSOP2) propõe uma abordagem prática e sequencial para o diagnóstico da sarcopenia, centrando-se inicialmente na deteção do risco através da velocidade da marcha. Uma velocidade de marcha inferior a 0,8 m/s sugere sarcopénia, a que se deve seguir a medição da força de preensão e a avaliação da massa muscular para confirmar o diagnóstico. A força de preensão é considerada baixa quando é inferior a 27 kg nos homens e a 16 kg nas mulheres. A massa muscular, avaliada por métodos como a absorciometria de raios X de dupla energia (DEXA) ou a bioimpedância eléctrica, também fornece critérios quantitativos para a confirmação do diagnóstico (Moretti et al.2024).

• Métodos e critérios de diagnóstico actuais

Para além do DEXA e da bioimpedância, podem ser utilizadas outras técnicas de imagiologia, como a tomografia computorizada (TC) e a ressonância magnética (RM), para uma avaliação pormenorizada da composição corporal. Estes métodos permitem não só identificar a redução da massa muscular, mas também analisar a qualidade do tecido muscular, incluindo a infiltração de gordura e a fibrose. A implementação destes critérios e métodos permite um diagnóstico preciso e diferenciado da sarcopénia, facilitando intervenções mais eficazes (Loyola et al.2020) (Criollo Sanchez, 2024).

• Abordagens terapêuticas e prevenção

A abordagem e tratamento da sarcopénia baseia-se numa combinação de intervenções farmacológicas e não farmacológicas. O tratamento baseia-se em 2 pilares principais: dietas ricas em proteínas (principalmente leucina e HMB) e treino de resistência e força, com o objetivo de promover o aumento da massa muscular e da força, devendo também ser administrados suplementos nutricionais em caso de carências proteicas e de vitamina D. É importante encorajar a atividade física regular, como caminhadas ou exercícios de equilíbrio, para melhorar a função muscular. Outras abordagens não farmacológicas incluem a terapia ocupacional para facilitar a independência nas actividades diárias e a fisioterapia para melhorar a mobilidade. Quanto às intervenções farmacológicas, não existe atualmente nenhuma estratégia eficaz para melhorar a massa ou a qualidade muscular, pelo que é importante avaliar os potenciais riscos e benefícios antes de iniciar qualquer tratamento farmacológico em doentes geriátricos (Barajas-Galindo et al.2021)(Murri, 2021).

f.Fragilidade

A síndrome de fragilidade é uma condição comum nos adultos mais velhos e caracteriza-se pela diminuição da reserva fisiológica e pelo aumento da vulnerabilidade ao stress. É definida como uma condição médica caracterizada pela presença de fraqueza, velocidade de marcha reduzida, diminuição da atividade física, perda de peso não intencional e diminuição da função cognitiva. A fragilidade afecta aproximadamente 10-25% das pessoas com mais de 65 anos e a prevalência aumenta com a idade. O diagnóstico de fragilidade baseia-se em critérios clínicos, que podem incluir a escala de fragilidade de Fried e o índice de fragilidade de Rockwood. A abordagem e o tratamento da fragilidade envolvem intervenções multidisciplinares, incluindo exercício físico, otimização da atividade física e terapia do exercício, nutrição, gestão da medicação, terapia cognitiva e redução dos factores de risco (Kaçmaz et al., 2023) (Montesino et al.2022). É essencial que os médicos residentes recebam formação sobre o reconhecimento e a gestão desta síndrome, a fim de prestarem cuidados de qualidade aos doentes geriátricos.

• Definição

A fragilidade é definida como um estado de maior vulnerabilidade a factores de stress externos, devido a uma diminuição das reservas fisiológicas do organismo.
múltiplos sistemas do corpo. Esta síndrome é caraterística da população idosa e esta associada a um risco elevado de eventos adversos, como a incapacidade, a hospitalização e a mortalidade. (Barillas Escobar & Henrfquez Mezquita, 2021).

• Epidemiologia

A síndrome de fragilidade é um problema geriátrico comum em todo o mundo, afectando especialmente pessoas com mais de 65 anos de idade. Estima-se que entre 4% e 27% dos idosos que vivem na comunidade e até 50% dos residentes em instituições geriátricas sofrem de fragilidade. Além disso, a prevalência da fragilidade aumenta com a idade sendo mais elevada nas mulheres (Martínez et al., 2024).

A fragilidade está também associada a um aumento da morbilidade e da mortalidade, bem como a uma maior utilização dos cuidados de saúde. Estes dados demonstram a importância de abordar e tratar adequadamente esta síndrome nos médicos residentes, a fim de prestar cuidados abrangentes e melhorar a qualidade de vida dos doentes geriátricos (Martfnez et al., 2024).

• Conceptualização e relevância em Geriatria

Em geriatria, a fragilidade é reconhecida como um preditor chave dos resultados de saúde em adultos mais velhos e um fator crítico no planeamento dos seus cuidados. A sua relevância reside na capacidade de identificar indivíduos com elevado risco de deterioração rápida, permitindo a implementação de estratégias preventivas e terapêuticas específicas que podem melhorar significativamente o seu prognóstico e qualidade de vida.

• Modelos de avaliação da fragilidade (G6mez Monedero, 2023) (L6pez, 2023)
Existem vários modelos de avaliação da fragilidade, dos quais se destacam dois:

- Modelo fenotípico de Fried: Este modelo identifica a fragilidade pela presença de cinco critérios: perda de peso não intencional, fraqueza (medida pela força de preensão), exaustão, lentidão e baixo nível de atividade física. A presença de três ou mais destes critérios classifica um indivíduo como frágil.

- Índice cumulativo de fragilidade: Esta abordagem considera a fragilidade como uma acumulação de défices ao longo do tempo, avaliando uma vasta gama de variáveis, incluindo sintomas, sinais, doenças, incapacidades e resultados laboratoriais. Um maior número de défices está associado a uma fragilidade mais grave.

• Impacto da fragilidade

O impacto da fragilidade na população idosa é profundo, afectando não só a saúde física, mas também a autonomia e a interação social. Os indivíduos frágeis são mais susceptíveis de sofrer de uma diminuição da qualidade de vida, de uma maior utilização dos serviços de saúde e de uma maior dependência de cuidados prolongados. Além disso, a fragilidade pode exacerbar o curso de outras doenças crónicas, complicando a sua gestão e tratamento.

• Intervenções para aumentar a resiliência e a qualidade de vida

A abordagem e o tratamento da síndrome de fragilidade em doentes geriátricos centram-se em cuidados abrangentes e multidimensionais.

É essencial conceber um plano de cuidados que aborde os factores de risco e as causas subjacentes à fragilidade. Isto implica uma abordagem interdisciplinar que inclua acções específicas para melhorar o funcionamento físico, nutricional e cognitivo dos doentes.

As intervenções não farmacológicas incluem o exercício físico personalizado, a terapia ocupacional, a estimulação cognitiva e uma alimentação adequada. Em termos de tratamento farmacológico, podem ser utilizados diferentes medicamentos para gerir as comorbilidades e os sintomas associados à fragilidade. Além disso, é essencial proporcionar um ambiente seguro e adaptado às necessidades do doente, bem como apoio social e emocional para melhorar a qualidade de vida. As intervenções para gerir a fragilidade centram-se na melhoria da resiliência e da capacidade funcional do indivíduo. Estas incluem:

- Programas de exercício físico: Concebidos para melhorar a força, o equilíbrio e a resistência geral.
- Otimização nutricional: Assegurar uma alimentação rica em proteínas, vitaminas e minerais essenciais à manutenção da massa muscular e da função imunitária.
- Gestão integrada das doenças crónicas: Coordenação dos cuidados entre diferentes especialistas para otimizar o tratamento das co-morbilidades.
- Apoio psicossocial: Prestação de assistência social e psicológica para promover a inclusão social e combater a depressão e a solidão.

3. Abordagem multidisciplinar das síndromes geriátricas

A abordagem multidisciplinar das síndromes geriátricas é essencial para prestar cuidados abrangentes aos doentes. Esta abordagem envolve uma variedade de profissionais de saúde, como médicos geriatras, enfermeiros, fisioterapeutas, terapeutas ocupacionais e assistentes sociais, entre outros. Cada profissional tem uma experiência e conhecimentos específicos para avaliar e tratar as síndromes geriátricas de diferentes perspectivas. O trabalho em equipa permite uma avaliação abrangente dos doentes, considerando tanto as dimensões médicas como as sociais, psicológicas e funcionais. É estabelecida uma comunicação eficaz entre os diferentes especialistas envolvidos, o que facilita a tomada de decisões partilhadas e a coordenação do tratamento. Isto melhora a qualidade de vida dos doentes geriátricos e reduz as complicações e as hospitalizações. Esta abordagem multidisciplinar é especialmente relevante nas síndromes geriátricas, uma vez que estas são frequentemente multifactoriais e requerem uma abordagem holística que considere todas as dimensões da pessoa idosa. O trabalho em equipa também permite uma visão mais abrangente da saúde do doente, uma vez que cada profissional pode detetar aspectos que outros podem não ver, resultando num tratamento mais eficaz e personalizado para cada indivíduo. A colaboração multidisciplinar proporciona um apoio abrangente aos doentes e às suas famílias, abordando não só as necessidades médicas, mas também as emocionais, sociais e de estilo de vida. Em suma, a abordagem multidisciplinar no tratamento das síndromes geriátricas representa uma forma holística e eficaz de melhorar a qualidade de vida e o bem-estar das pessoas idosas.

4. Tratamento farmacológico e não-farmacológico Tratamento farmacológico

O tratamento das síndromes geriátricas é abordado de forma farmacológica e não farmacológica. No caso do tratamento farmacológico, os medicamentos são utilizados para tratar os sintomas e as causas subjacentes das síndromes geriátricas. Por exemplo, na síndrome de imobilidade, podem ser prescritos medicamentos para a dor e a inflamação, bem como para melhorar a mobilidade. No caso do tratamento não farmacológico, são utilizadas intervenções e terapias não medicamentosas. Estas podem incluir terapia física e ocupacional, programas de exercício, alterações alimentares e adaptação do ambiente físico. Podem também ser utilizadas técnicas de terapia cognitiva e emocional para tratar síndromas geriátricos como o défice cognitivo e a fragilidade. Em geral, os tratamentos farmacológicos e não farmacológicos são utilizados de forma complementar para otimizar a gestão das síndromes geriátricas e melhorar a qualidade de vida dos doentes idosos.

5. Importância da Formação em Síndromes Geriátricas para Médicos Residentes

A formação em síndromes geriátricas é da maior importância para os médicos residentes devido ao envelhecimento crescente da população e à crescente procura de cuidados médicos neste grupo de doentes.

As síndromes geriátricas são condições clínicas comuns em adultos mais velhos e a sua gestão adequada requer conhecimentos específicos e competências clínicas. Estas síndromes, como a imobilidade, o café, o défice cognitivo, a sarcopenia e a fragilidade estão associadas a um risco acrescido de incapacidade, hospitalização, institucionalização e mortalidade.

A falta de formação adequada em síndromes geriátricas pode levar a um diagnóstico e tratamento inadequados, bem como a um aumento das complicações e a resultados negativos para os doentes. Por conseguinte, é essencial que os médicos residentes adquiram os conhecimentos e as competências necessárias para identificar e gerir estas síndromes de uma forma abrangente, prestando cuidados óptimos e de qualidade aos adultos mais velhos.

Deve ser dada especial atenção à importância da gestão adequada dos medicamentos nos doentes idosos, uma vez que uma farmacoterapia inadequada pode aumentar o risco de interacções medicamentosas e de efeitos adversos. É essencial promover a educação continuada em síndromes geriátricas para garantir que os médicos residentes sejam treinados para fornecer cuidados abrangentes e de qualidade à crescente população de adultos mais velhos.

6. B-learning no Ensino das Síndromes Geriátricas

O B-learning é uma metodologia inovadora que combina eficazmente a aprendizagem presencial e online para facilitar o ensino de Sintomas Geriátricos aos médicos residentes. Esta estratégia educativa baseia-se na utilização de recursos digitais e tecnológicos que permitem aos alunos aceder a conteúdos teóricos, estudos de caso, vídeos e outros recursos educativos a partir de qualquer local e em qualquer altura. Incentiva a participação ativa através de fóruns de discussão, actividades colaborativas e avaliações online, promovendo a aquisição de conhecimentos teóricos e práticos, bem como o desenvolvimento de competências clínicas e a tomada de decisões informadas na abordagem e tratamento dos Sintomas Geriátricos.Ao combinar o ensino presencial e online, o B-Learning oferece flexibilidade e adaptabilidade às necessidades individuais dos médicos residentes, permitindo-lhes aceder aos conteúdos de forma autónoma e desenvolver a sua aprendizagem ao seu próprio ritmo. Esta metodologia proporciona uma experiência de aprendizagem mais enriquecedora e eficaz, melhorando a retenção de conhecimentos e a transferência de competências para a prática clínica. Em resumo, o B-learning é uma estratégia educativa inovadora que optimiza o ensino da Sintomatologia Geriátrica aos médicos residentes, fornecendo-lhes ferramentas e recursos de aprendizagem de última geração para melhorar a sua formação e desempenho profissional. O método B-learning ajuda os médicosO programa foi concebido para permitir que os médicos residentes adquiram competências técnicas e conhecimentos teóricos de uma forma flexível e adaptável às suas necessidades individuais. Esta metodologia inovadora é altamente eficaz e enriquecedora, permitindo que os médicos residentes desenvolvam as suas competências ao seu próprio ritmo e em qualquer local. A abordagem B-learning incentiva a participação ativa através de fóruns de discussão e actividades de colaboração, permitindo aos estudantes aprender de uma forma mais holística. Graças à combinação da aprendizagem presencial e em linha, os médicos residentes podem melhorar significativamente a sua formação e desempenho profissional.

a. Metodologias e ferramentas de B-learning

No ensino de Sintomas Geriátricos a médicos residentes através do B-learning, podem ser utilizadas várias metodologias e ferramentas para apoiar o processo de ensino-aprendizagem. As metodologias incluem uma combinação de aulas presenciais e virtuais, permitindo flexibilidade de tempo e espaço para a aprendizagem. Além disso, são utilizadas técnicas como estudos de casos clínicos, simulações virtuais e debates, incentivando a participação ativa dos alunos e a aplicação dos conhecimentos teóricos em situações reais. Estas metodologias e ferramentas de B-learning contribuem para melhorar a formação dos médicos residentes em doenças geriátricas, promovendo uma aprendizagem mais dinâmica, participativa e centrada no aluno.

b. Conceção de conteúdos para o ensino de síndromes geriátricas em B-learning

A conceção dos conteúdos para o ensino das síndromes geriátricas em b-learning é essencial para garantir uma aprendizagem eficaz. Os objectivos de cada módulo, bem como os conteúdos a abordar, devem ser definidos de forma clara e concisa. Deve ter-se em conta que o B-learning combina o ensino presencial com a aprendizagem online, pelo que é necessário adaptar os materiais e as estratégias de ensino a esta modalidade. Para o efeito, podem ser utilizados vários recursos, como apresentações interactivas, vídeos educativos, estudos de caso e fóruns de discussão. É essencial estabelecer uma sequência lógica e coerente na apresentação dos conteúdos para que os médicos residentes possam adquirir conhecimentos de forma progressiva. Devem ser incluídas actividades de avaliação para medir o nível de compreensão e de retenção da informação. Em resumo, a conceção de conteúdos para o ensino das síndromes geriátricas em B-learning deve ser didática, interactiva e adaptada às necessidades dos médicos residentes.

i. Metodologias de conceção

No ensino de sintomas geriátricos a médicos residentes através do B-learning, as metodologias de conceção desempenham um papel fundamental.

1. Fóruns virtuais

Os fóruns virtuais são uma forma de interação em linha que permite aos médicos residentes discutir e partilhar ideias sobre síndromes geriátricas. Estes fóruns proporcionam um espaço virtual onde os participantes podem colocar questões, fazer e responder a comentários e partilhar recursos relevantes (Fernandez et al.2022). Os fóruns virtuais incentivam a participação ativa e a partilha de conhecimentos entre médicos residentes, o que facilita a aprendizagem colaborativa e o desenvolvimento de competências na gestão e tratamento de síndromes geriátricas. Neste sentido, os fóruns virtuais são uma ferramenta eficaz para promover a discussão e a aprendizagem entre médicos residentes no contexto do B-learning; são uma ferramenta essencial para a troca de informações e a colaboração entre médicos de diferentes especialidades, o que contribui significativamente para o avanço da medicina geriátrica. A possibilidade de conversas fluidas através de plataformas virtuais permite que os profissionais de saúde acedam a diferentes perspectivas, descubram novas práticas e melhorem as suas competências clínicas, a fim de prestarem os melhores cuidados aos doentes geriátricos. A flexibilidade e a facilidade de acesso aos fóruns virtuais permitem que os médicos se mantenham actualizados sobre temas relevantes e participem no desenvolvimento de novas investigações e na implementação de melhores práticas no tratamento de doentes idosos. (Céspedes-Tamayo et al.2020).

Em suma, os fóruns virtuais representam uma ferramenta valiosa para o reforço da comunidade médica e para a melhoria contínua dos cuidados geriátricos.

2. Simulação virtual

A simulação virtual é uma metodologia utilizada no ensino de doenças geriátricas aos médicos residentes. A simulação virtual envolve a utilização de software e tecnologia de realidade virtual para recriar situações clínicas realistas e permitir que os estudantes pratiquem e adquiram competências num ambiente seguro e controlado. Através da simulação virtual, os médicos residentes podem ser confrontados com casos clínicos complexos e aplicar os seus conhecimentos teóricos na tomada de decisões médicas. A simulação virtual oferece a vantagem de ser flexível em termos de tempo e local de estudo, uma vez que pode ser feita individualmente e em momentos diferentes.

Esta metodologia de ensino é particularmente útil para a aquisição de aptidões clínicas práticas e para o desenvolvimento das competências necessárias à gestão e ao tratamento das síndromas geriátricas. A utilização da simulação virtual permite aos médicos residentes melhorar a sua capacidade de diagnosticar e tratar eficazmente os doentes geriátricos. Ao proporcionar um ambiente realista e seguro, os médicos residentes podem praticar em cenários que imitam a complexidade e os desafios dos cuidados médicos prestados aos idosos. Isto dá-lhes a oportunidade de aplicar os seus conhecimentos num ambiente controlado e receber feedback imediato sobre as suas decisões. (Siew et al., 2021).

A simulação virtual também lhes permite aprender a gerir situações clínicas de grande complexidade, o que é crucial para a sua formação como médicos. Além disso, ao poderem efetuar a simulação em horários e locais flexíveis, os médicos residentes podem adaptar a sua formação às suas próprias necessidades e responsabilidades (Chaby et al.2022). Em suma, a simulação virtual é uma ferramenta valiosa para a formação de médicos residentes na gestão de síndromes geriátricas, proporcionando uma forma segura, eficaz e flexível de adquirir e melhorar as competências clínicas.

3. Sala de aula invertida

A sala de aula invertida é uma metodologia de ensino que promove a aprendizagem ativa e a participação dos médicos residentes. Nesta modalidade, os alunos estudam os conteúdos teóricos das síndromes geriátricas de forma autónoma antes da aula, utilizando materiais didácticos como vídeos, leituras ou tutoriais online. Posteriormente, em sala de aula, são incentivadas a discussão, a análise de casos e a resolução de problemas práticos, o que facilita a aplicação dos conhecimentos adquiridos. Esta estratégia permite uma melhor utilização do tempo de aula e promove uma aprendizagem mais profunda, uma vez que os médicos residentes podem resolver dúvidas e receber feedback direto do professor. Ao realizar actividades práticas em grupo, incentiva-se o trabalho em equipa e a colaboração entre os participantes, competências essenciais na gestão das síndromes geriátricas. A sala de aula invertida cria um ambiente de aprendizagem colaborativa que permite aos residentes desenvolverem competências cognitivas complexas e críticas. A construção do conhecimento é incentivada através do debate e da discussão, permitindo aos residentes analisar e refletir sobre o material de aprendizagem a partir de diferentes perspectivas. O modelo também incentiva a responsabilidade individual dos residentes no processo de aprendizagem, promovendo a preparação prévia e a participação ativa nas actividades da sala de aula. Esta metodologia também promove o desenvolvimento de competências de comunicação eficazes, ensinando os residentes a exprimir as suas ideias de forma clara e coerente. (Masud et al 2022)(Ong et al.2021)(Wu et al., 2020)

Em suma, a sala de aula invertida é uma técnica eficaz para ensinar sobre estas síndromes, uma vez que incentiva a aprendizagem ativa, a participação e o trabalho de equipa dos médicos residentes. Em suma, a sala de aula invertida é uma ferramenta poderosa para melhorar a qualidade do ensino médico através da participação ativa dos médicos residentes e da promoção de uma aprendizagem significativa e colaborativa. A sala de aula invertida enriquece a experiência educativa ao promover uma discussão construtiva que permite aos médicos residentes analisar e refletir sobre as síndromes geriátricas a partir de uma variedade de perspectivas. Além disso, o modelo fomenta a investigação crítica e autónoma, incentivando os residentes a envolverem-se ativamente no seu processo de aprendizagem e a procurarem ativamente a compreensão e o domínio dos conteúdos.

4. Classe assíncrona

A sala de aula assíncrona é uma metodologia de ensino utilizada no B-learning que permite aos médicos residentes aceder a materiais e conteúdos de forma flexível e autónoma. Nesta modalidade, os participantes podem aceder a vídeos, leituras e outros recursos online ao seu próprio ritmo, dando-lhes a oportunidade de rever conteúdos e aprofundar temas de interesse. (Incentiva a interação entre os médicos residentes através de fóruns virtuais e outras ferramentas de comunicação em linha, onde podem discutir e resolver questões. Esta abordagem permite que os residentes adaptem a sua aprendizagem às suas próprias necessidades e horários, o que promove uma maior participação e compreensão no estudo das síndromes geriátricas. A flexibilidade da sala de aula assíncrona também facilita o acesso à educação médica contínua, pois permite que os residentes continuem a aprender apesar das obrigações profissionais e pessoais. (Ong et al.2021)Em suma, esta metodologia incentiva a autonomia e a responsabilidade no processo de aprendizagem preparando os médicos residentes para uma prática profissional mais reflexiva e comprometida com a atualização constante dos conhecimentos médicos.

5. Explicações virtuais

A tutoria virtual é uma ferramenta fundamental no B-learning das síndromes geriátricas. Durante as tutorias virtuais, os médicos residentes podem colocar questões, receber feedback sobre os seus casos clínicos e discutir estratégias de abordagem e tratamento. A tutoria virtual favorece a aprendizagem autónoma e a interação entre tutores e residentes, promovendo um espaço de reflexão e análise de casos reais. Da mesma forma, este tipo de tutoria facilita o acesso a informação actualizada e a recursos bibliográficos relevantes (Masud et al.2022).

A tutoria virtual também lhes permite desenvolver competências de comunicação e colaboração através de plataformas virtuais, o que é essencial no ambiente atual dos cuidados de saúde. Além disso, ao promover a autonomia dos residentes, a tutoria virtual dá-lhes um maior sentido de responsabilidade e controlo sobre a sua aprendizagem. Os tutores especializados podem partilhar experiências clínicas e estudos de caso relevantes, o que enriquece a formação dos residentes e lhes dá uma perspetiva prática que complementa a sua formação clínica (Ong et al.2021)(Pan et al., 2024). Através da tutoria virtual, é possível aceder a sessões de formação especializadas e a palestras de líderes de opinião no domínio da geriatria. Em suma, a tutoria virtual é uma estratégia pedagógica eficaz para reforçar os conhecimentos e as competências dos médicos residentes na gestão de síndromes geriátricas através da interação direta com especialistas na área, que promove o desenvolvimento profissional dos residentes e os ajuda a manterem-se actualizados com as últimas tendências e avanços no tratamento de síndromes geriátricas.

c. Experiências no ensino médico

No domínio da educação médica, têm sido desenvolvidas várias experiências no ensino das síndromes geriátricas através da utilização da modalidade B-learning. Estas experiências têm-se centrado em proporcionar aos médicos residentes a oportunidade de alargarem os seus conhecimentos sobre as diferentes síndromes geriátricas, bem como de lhes fornecerem ferramentas práticas para a sua abordagem e tratamento, permitindo-lhes adquirir as aptidões e competências necessárias para prestarem cuidados integrais e de qualidade aos doentes geriátricos. (No entanto, também foram identificadas limitações na implementação do B-learning no ensino médico, tais como a falta de acesso a recursos tecnológicos, a resistência à mudança ou a dificuldade em avaliar a aprendizagem em ambientes virtuais. (Apesar destes desafios, experiências bem sucedidas no ensino de doenças geriátricas com uma abordagem B-learning representam uma oportunidade para melhorar a formação dos médicos residentes e garantir cuidados adequados à população geriátrica.
i. Expectativas e limitações

Relativamente às expectativas do B-learning no ensino das síndromes geriátricas aos médicos residentes, espera-se que esta metodologia proporcione um ambiente flexível e acessível para a aprendizagem, permitindo aos alunos aceder aos conteúdos em qualquer altura e lugar.

Espera-se que a utilização de ferramentas virtuais e a interação com outros estudantes e tutores em fóruns virtuais promovam um maior envolvimento e participação ativa no processo de aprendizagem. No entanto, é importante ter em conta as limitações desta metodologia, como a necessidade de acesso à Internet e a dispositivos tecnológicos, bem como a possibilidade de alguns alunos terem dificuldades de adaptação a esta forma de aprendizagem. Deve também ter-se em conta que o B-learning requer um planeamento cuidadoso para garantir a integração eficaz das diferentes componentes e actividades de aprendizagem.

d. Avaliação da Aprendizagem em Síndromes Geriátricas com uma Abordagem B-learning

A avaliação da aprendizagem em doenças geriátricas com uma abordagem B-learning é essencial para medir o conhecimento adquirido pelos médicos residentes. Devem ser utilizadas diferentes estratégias de avaliação, tais como testes escritos, estudos de caso, actividades práticas e avaliação contínua. Para além disso, é importante avaliar o processo de aprendizagem em síndromes geriátricas, identificando os pontos fortes e fracos dos médicos residentes, de modo a fornecer feedback adequado e a melhorar continuamente o ensino.

A avaliação deve ser objetiva, justa e baseada em critérios estabelecidos, a fim de garantir a qualidade e o rigor académico na formação dos médicos residentes.

e. Experiências de sucesso e boas experiências

Prática no Ensino de Síndromes Geriátricas com B-learning No ensino de Sintomas Geriátricos com B-learning, foram identificadas várias experiências de sucesso e boas práticas. Algumas delas incluem a integração de casos clínicos reais e relevantes para que os médicos residentes possam aplicar os seus conhecimentos teóricos em situações práticas.

A utilização de plataformas virtuais interactivas, onde os residentes podem aceder a materiais de estudo, realizar actividades e participar em fóruns de discussão com outros profissionais, também se revelou benéfica (Siew et al., 2021).

É importante ter uma equipa de ensino multidisciplinar que forneça apoio personalizado e feedback aos residentes (Masud et al.2022). Estas experiências de sucesso e boas práticas permitiram melhorar a aprendizagem e formação dos médicos residentes na abordagem e tratamento das Síndromes Geriátricas a partir da abordagem B-learning.

f.Desafios e Oportunidades na Implementação do B-Learning no Ensino das Síndromes Geriátricas

A implementação do B-learning no ensino de doenças geriátricas apresenta uma série de desafios e oportunidades.

Um dos desafios é a adaptação de conteúdos e metodologias tradicionais a um formato de aprendizagem B-learning, que combina a aprendizagem presencial e em linha. Para tal, é necessário rever e ajustar os materiais didácticos para garantir que são acessíveis e eficazes tanto em ambientes físicos como virtuais.

Os educadores precisam de ser formados na utilização das ferramentas e dos recursos tecnológicos necessários para implementar com êxito a aprendizagem eletrónica. A implementação do B-learning também oferece oportunidades significativas. Através do B-learning, os médicos residentes podem aceder a uma variedade de recursos online, tais como vídeos, estudos de caso e simulações, que enriquecem o seu processo de aprendizagem. A aprendizagem online permite a adaptação do tempo e do ritmo de estudo de cada residente, potenciando a sua autonomia e auto-disciplina.

Em resumo, a implementação do B-learning no ensino de doenças geriátricas coloca desafios em termos de adaptação e formação, mas também oferece oportunidades para melhorar a acessibilidade e a qualidade da aprendizagem para os médicos residentes.

7. Conclusões e Recomendações para o Ensino de Síndromes Geriátricas a Médicos Residentes

Em conclusão, o ensino das síndromes geriátricas aos médicos residentes é de importância vital para preparar os profissionais de saúde para o tratamento integral da população idosa.

É essencial que os médicos residentes adquiram um conhecimento sólido sobre a definição, os conceitos, a epidemiologia, as características, a abordagem e o tratamento dos diferentes síndromes geriátricos, tais como a imobilidade, a coabitação, o défice cognitivo, a sarcopenia e a fragilidade. Isto permitir-lhes-á oferecer cuidados de qualidade, baseados em provas científicas e adaptados às necessidades específicas dos doentes geriátricos. Por conseguinte, recomenda-se a inclusão de conteúdos específicos sobre síndromes geriátricas nos programas de formação dos médicos residentes, bem como a utilização de metodologias e ferramentas B-learning para otimizar o processo de ensino-aprendizagem.

É essencial encorajar uma abordagem multidisciplinar e a utilização de abordagens farmacológicas e não farmacológicas no tratamento destas síndromes. Considero importante a partilha de experiências de sucesso e de boas práticas entre os profissionais de saúde, de forma a gerar conhecimento e a melhorar a qualidade do ensino nesta área. Apesar dos desafios e oportunidades apresentados pela implementação do B-learning no ensino das doenças geriátricas, é essencial continuar a progredir nesta direção para formar médicos residentes competentes e empenhados no cuidado da população geriátrica.

REFERÊNCIAS

• Pefia, K. P., Morera, M. R., Chaves, F. O., Quir6s, K. V. L., & Quir6s, S. L. (2020). Síndromes geriátricas: café, incontinência e comprometimento cognitivo. Revista Hispanoamericana de Ciencias de la Salud (RHCS), 6(4), 201-210. unirioja.es

• Corzo Camacho, M. A. (). Construcci6n de un m6dulo educativo para la ensefianza de los grandes sfndromes geriatricos por medio del uso de aprendizaje basado en problemas y tecnologfas .. repositorio.unal.edu.co. unal.edu.co.

• Tamayo Pérez, L. (2023). Fragilidade e síndromes geriátricos num grupo de idosos institucionalizados. uva.es

• Suarez Tomala, G. A. (2022). Deterioração da mobilidade física e sua influência no bem-estar psicológico dos idosos. Centro de saúde Basti6n Popular tipo C. Guayaquil, 2022. upse.edu.ec

• Cuasapaz-Bermeo, A. E., Davas-Torres, C. C., Granda-Carbo, M. V., Zambrano-Santana, J. P., & Ponce-Alencastro, J. A. (2023). Avaliação clínica e acompanhamento de úlceras de pressão em pacientes geriátricos: Uma revisão integrada da literatura. Multidisciplinary & Health Education Journal, 5(2), 279-285. journalmhe.org.

• MUNOZ, D. Y. S. (). . ESTUDO DA RELAÇÃO DE VARIÁVEIS SOCIODEMOGRÁFICAS, CLÍNICAS E DE CAPACIDADE FUNCIONAL COM O RISCO DE QUEDAS EM ADULTOS. Presidentes de Câmara do Município de ... core.ac.uk. core.ac.uk

• Sanchez, A. (2023). Abordaje kinésico domiciliario en patologfas del adulto mayor. ufasta.edu.ar

• G6mez Monedero, A. (2023). Aproximaci6n a los instrumentos de evaluación del sfundrome de fragilidad: scoping review. universidadeuropea.com.

• Cedefio, K. O. C., Narvaez, E. R. C., Contreras, J. N. I., & Ortiz, B. D. G. (2024). Polineuropsicofarmácia: Revisão atualizada de suas complicações na população geriátrica. Ciencia Latina Revista Cientffica Multidisciplinar, 8(1), 1759-1775. ciencialatina.org.

• Franco, F. & Lisbeth, G. (2022). Factores de riesgo de cafda en los adultos mayores del barrio Parafso, parroquia Anconcito 2022... upse.edu.ec

• Hart, L. A., Phelan, E. A., Yi, J. Y., Marcum, Z. A., & Gray, S. L. (2020). Uso de drogas que aumentam o risco de queda em torno de uma lesão relacionada à queda em adultos mais velhos: Uma revisão sistemática. Journal of the American Geriatrics Society, 68(6), 1334- 1343. nih.gov.

• Ganz, D. A. & Latham, N. K. (2020). Prevenção de quedas na comunidade-idosos que vivem em casa. New England journal of medicine. escholarship.org

• Iglesias, A. L., Sanchez, L. H., Mateos-Nozal, J., & Nebreda, M. Â. (2022). Cafdas e fratura de quadril. Medicine-Programa de Formaci6n Médica Continuada Acreditado, 13(62), 3659-3670.

• COLCHADO ROSALES, B. (2021). Efectividad de ejercicios de coordinación y equilibrio en la marcha del adulto mayor en un Hospital Publico, Chimbote 2019. usanpedro.edu.pe

• Chorefio-Parra, J. A., De la Rosa-Arredondo, T., & Guadarrama-Ortfz, P. (2020). Abordaje diagn6stico del paciente con deterioro cognitivo en el primer nivel de atenci6n. Med Int Méx, 36(6). utel.edu.mx

• Parada Mufioz, K. R., Guapizaca Juca, J. F., & Bueno Pacheco, G. A. (2022). Comprometimento cognitivo e depressão em idosos: uma revisão sistemática dos últimos 5 anos. Revista Cientffica UISRAEL, 9(2), 77-93. senescyt.gob.ec

• Fonte Sevillano, T. & Santos Hedman, D. J. (2020). Comprometimento cognitivo leve em pessoas com mais de 85 anos. Revista cubana de medicina. sld.cu.

• Galvez, C. M. G. (2024). Utilidade do teste STROOP na avaliação das funções cognitivas em pessoas com demência. CIENCIAMATRIA. unirioja.es

• Valenzuela Sanchez, E. J. & Marcelino Arias, N. K. (2023). . Correlação entre o comprometimento cognitivo identificado no teste de mini-exame do estado mental e os critérios diagnósticos para a doença de Alzheimer no hospital.unphu.edu.do

• Quinaloa, J. G. L., Guamangate, Y. K. M., Caisaluisa, J. L. M., & Cerda, V. D. C. T. (2020). Teste minimental para o diagnóstico precoce do défice cognitivo. Revista de Investigação INNOVA, 5(3), 13. unirioja.es

• Cancino, M., Rehbein, L., G6mez-Pérez, D., & Ortiz, M. S. (2020). Avaliação do funcionamento cognitivo em adultos: Análise e contraste de três dos instrumentos mais utilizados no Chile. Revista médica de Chile, 148(4), 452-458. scielo.cl

• Sailema Lalaleo, A. J. (2023). Gufa de estimulaci6n cognitiva y su efecto en adultos mayores con deterioro cognitivo. uta.edu.ec

• Mazo Bafiol, Y. & Moncada Botero, M. (2023). Protocolo de estimulaci6n cognitiva de las funciones ejecutivas en adultos mayores con demencia por cuerpos de Lewy. ces.edu.co

• Garcfa-Ribas, G., Marfn, A. S., & Barreto, P. L. (2023). Tratamento do défice cognitivo. Medicine-Programa de Formaci6n Médica Continuada Acreditado, 13(74), 4382-4394.

• Radoja, I., & Degmecié, D. (2020). Incontinência urinária: avaliação diagnóstica e tratamento de primeira linha. Jornal Médico do Sudeste Europeu: SEEMEDJ, 4(1), 63-73. srce.hr

• Nambiar, A. K., Arlandis, S., B0, K., Cobussen-Boekhorst, H., Costantini, E., de Heide, M., ... & Harding, C. K. (2022). Diretrizes da associação europeia de urologia sobre o diagnóstico e tratamento dos sintomas femininos não neurogênicos do trato urinário inferior. Parte 1: diagnóstico, bexiga hiperactiva, incontinência urinária de esforço e incontinência urinária mista. European Urology, 82(1), 49-59. abdn.ac.uk

• Shaw, C. & Wagg, A. (2021). Incontinência urinária e fecal em adultos mais velhos. Medicina. [HTML].

• O'Connor, E., Nic an Riogh, A., Karavitakis, M., Monagas, S., & Nambiar, A. (2021). Diagnóstico e tratamento não cirúrgico da incontinência urinária - uma revisão da literatura com recomendações para a prática. International Journal of General Medicine, 4555-4565. tandfonline.com.

• Pérez Carvajal, G. J. (2023). Relaci6n entre ingesta de protefna de alto valor biol6gico y prevalencia de Sarcopenia en adultos mayores de un centro geriatrico de la provincia de

Chimborazo .. espoch.edu.ec

• Sandoval Animas, G. E. (). Sarcopenia y malnutrición en personas mayores, una revisión bibliografica actualizada. repositorio.xoc.uam.mx. uam.mx

• Petermann-Rocha, F., Balntzi, V., Gray, S. R., Lara, J., Ho, F. K., Pell, J. P., & Celis-Morales, C. (2022). Prevalência global de sarcopenia e sarcopenia grave: uma revisão sistemática e meta-análise. Journal of cachexia, sarcopenia and muscle, 13(1), 86-99. wiley.com.

• Gao, Q., Mei, F., Shang, Y., Hu, K., Chen, F., Zhao, L., & Ma, B. (2021). Global prevalência de obesidade sarcopénica em adultos mais velhos: Uma revisão sistemática e meta-análise. Nutrição Clínica. researchgate.net

• Almohaisen, N., Gittins, M., Todd, C., Sremanakova, J., Sowerbutts, A. M., Aldossari, A., ... & Burden, S. (2022). Prevalência de desnutrição, fragilidade e sarcopenia em pessoas residentes na comunidade com 50 anos ou mais: revisão sistemática e meta-análise. Nutrientes, 14(8), 1537. mdpi.com

• Moretti, D., Fiorillo, P., Mogliani, M., Buncuga, M., & Fain, H. (2024). Avaliação da sarcopenia e dos parâmetros de bioimpedância relacionados à força muscular na consulta pré-operatória de cirurgia da coluna vertebral. Nutrición Hospitalaria, 41(1), 145-151. isciii.es

• Loyola, W. A. S., Corrales, G. A. L., Ganz, F., Caro, H. G., & Probst, V. S. (2020). Sarcopenia, definição e diagnóstico: precisamos de valores de referência para idosos na América Latina? Revista Chilena de Terapia Ocupacional, 20(2), 259-267. researchgate.net

• Criollo Sanchez, J. I. (2024). Uso da dinamometria em adultos mais velhos para determinar a sarcopenia. Revisión narrativa. udla.edu.ec

• Acosta-Benito, M. & Martín-Lesende, I. (2022). Fragilidade nos cuidados primários: Diagnóstico e gestão multidisciplinar. Atención Primaria. sciencedirect.com

• Barajas-Galindo, D. E., Arnaiz, E. G., Vicente, P. F., & Ballesteros-Pomar, M. D. (2021). Efeitos do exercício físico em idosos com sarcopenia. Uma revisão sistemática. Endocrinologia, Diabetes e Nutrição, 68(3), 159-169. [HTML]

• Murri, M. (2021). Efeitos da cinesioterapia na sarcopenia em idosos. ugr.edu.ar

• Castro Ellis, A. & Córdoba Granados, J. (). Beneficios no cardiovasculares del ejercicio ffsico en adultos mayores. kerwa.ucr.ac.cr. ucr.ac.cr

• Kaçmaz, H. Y., Doner, A., Kahraman, H., & Akin, S. (2023). Prevalência e fatores associados à fragilidade em pacientes idosos hospitalizados. Revista Clfnica Espafiola. unirioja.es

• Montesino, D. C., Reguera, I. P., Fernandez, O. R., Relova, M. R., & Valladares, W. C. (2022). Caracterização clínica e epidemiológica da deficiência na população idosa. Interdisciplinary Rehabilitation/Rehabilitacion Interdisciplinaria, 2, 15-15. saludcyt.ar

• Raymundo, R. R., Marfa, C. C. R., Edith, M. E. D., & Ernesto, R. O. R. (). Alteración de la velocidad de la marcha y del test de levantarse de la silla:L Inicio del sfrome de fragilidad en mujeres mayores institucionalizadas?. academia.edu. academia.edu

• Barillas Escobar, E. J. & Henrfquez Mezquita, M. E. (2021). Analisis de la aplicación de la Escala de Fried en el diagnostico de fragilidad en el adulto mayor que consulta a la Clfnica

Comunal San Antonio Abad de la Redues.edu.sv

• Martfnez, J. M. O., Martfnez, P. H., & Macfas, J. G. (2024). Fragilidade, sarcopenia e osteoporose. Medicina Clfnica. [HTML].

• L6pez, T. E. (2023). Construcci6n, disefio y validaci6n de un instrumento de evaluación6n sobre el conocimiento del sfrome de fragilidad en adultos mayores. uaq.mx

• Fernandez, A. M., Reyes, M. J., & L6pez, M. I. V. (2022). As tecnologias de informação e comunicação (TIC) na formação e no ensino. FMC- Formaci6n Médica Continuada en Atenci6n Primaria, 29(3), 28-38. [HTML] [HTML

• Sanchez, I. V. M. D. O., Bravo, M. G. E., Reyes, A. T. C., Marfn, H. J. V., & Chacha, A. G. O. (2023). EduTrends: Navegando na Era Digital da Educação. Editorial Investigativa Latinoamericana (SciELa). google.com

• Céspedes-Tamayo, L. G., Augello-Dfaz, S. L., & Ulloa-Cedefio, H. A. (2020). As redes sociais no processo de ensino-aprendizagem. XIII Jornada de Aprendizaje en Red. researchgate.net

• Chaby, L., Benamara, A., Pino, M., Prigent, E., Ravenet, B., Martin, J. C., ... & Chetouani, M. (2022). Pacientes virtuais incorporados como uma estrutura baseada em simulação para o treinamento de habilidades de comunicação clínico-paciente: Uma visão geral de seu uso em cuidados psiquiátricos e geriátricos. Frontiers in Virtual Reality, 3, 827312. frontiersin.org.

• Siew, A. L., Wong, J. W., & Chan, E. Y. (2021). Eficácia de pacientes simulados na educação geriátrica: Uma revisão de escopo. Educação em enfermagem hoje. [HTML].

• Masud, T., Ogliari, G., Lunt, E., Blundell, A., Gordon, A. L., Roller-Wirnsberger, R., ... & Stuck, A. E. (2022). A scoping review of the changing landscape of geriatric medicine in undergraduate medical education: curricula, topics and teaching methods. European geriatric medicine, 13(3), 513-528. springer.com.

• Ong, E. Y., Bower, K. J., & Ng, L. (2021). Intervenções educacionais geriátricas para médicos em treinamento em especialidades não geriátricas: uma revisão de escopo. Journal of Graduate Medical Education, 13(5), 654-665. allenpress.com.

• Wu, S., Jackson, N., Larson, S., & Ward, K. T. (2020). Ensino de Geriatria e Transições de Cuidados para Médicos Residentes de Medicina Interna. Geriatria. mdpi.com

• Pan, F., Ge, L., Hu, M., Liu, M., & Jiang, W. (2024). Aplicação do diagnóstico e tratamento virtual combinado com o método de ensino do registo médico na formação padronizada do médico de clínica geral. Medicina. lww.com

• Stefanowicz-Kocol, A., Grochowska, A., & Kolpa, M. (2023). Um modelo de ensino e aprendizagem de simulação mista / à distância sensível à cultura no campo da geriatria. atar.edu.pl

• Piot, M. A., Dechartres, A., Attoe, C., Jollant, F., Lemogne, C., Layat Burn, C., & Falissard, B. (2020). Simulação em psiquiatria para médicos: uma revisão sistemática e meta-análise. Educação médica, 54(8), 696-708. sorbonne-universite.fr

SEQUÊNCIAS DIDÁCTICAS PARA O ENSINO DAS SÍNDROMES GERIÁTRICAS AOS MÉDICOS RESIDENTES

Introdução

O envelhecimento da população é um fenómeno global que coloca desafios significativos e oportunidades únicas para o campo da medicina. À medida que aumenta a proporção de adultos mais velhos na sociedade, aumenta também a prevalência de síndromas geriátricos específicos que requerem cuidados especializados e uma gestão cuidadosa. Entre estas, destacam-se a fragilidade, a sarcopenia, o défice cognitivo, a anestesia e a incontinência urinária como condições críticas que afectam consideravelmente a qualidade de vida dos idosos. Estas síndromes não só afectam a saúde física e mental dos indivíduos, como também impõem encargos significativos aos prestadores de cuidados, aos sistemas de saúde e à sociedade em geral. Reconhecendo a necessidade crescente de uma formação eficaz e especializada em geriatria, este livro foi concebido especificamente para professores que estão na vanguarda da educação médica. O nosso objetivo é fornecer um guia completo para o ensino das síndromes geriátricas, utilizando uma abordagem de aprendizagem mista ou B-learning, que combina métodos de ensino tradicionais com tecnologias digitais modernas. Esta metodologia não só enriquece a experiência de aprendizagem, como também permite uma maior flexibilidade, acessibilidade e adaptabilidade, elementos essenciais para a formação de uma nova geração de médicos capazes de enfrentar os desafios da medicina geriátrica com competência e compaixão.Este capítulo fornece uma série de sequências de ensino pormenorizadas que abrangem cada uma das principais síndromes geriátricas. Estas sequências foram concebidas para facilitar aos professores o planeamento e a implementação de cursos eficazes, fornecendo estruturas claras, recursos educativos recomendados, actividades práticas e métodos de avaliação. Ao integrar a teoria e a prática, e ao fornecer numerosos exemplos de como aplicar os conhecimentos em situações clínicas da vida real, esperamos promover uma aprendizagem significativa e duradoura. Ao fornecer este guia abrangente, pretendemos apoiar os educadores nos seus esforços para preparar os futuros médicos, dotando-os das competências e conhecimentos necessários para melhorar os cuidados e o bem-estar da população idosa. Acreditamos firmemente que a educação médica de qualidade é a pedra angular para alcançar cuidados geriátricos excepcionais e responder eficazmente às necessidades de uma população envelhecida.

1. Sequência Didática Detalhada para a Síndrome de Imobilidade

Objetivo: Aprofundar a compreensão das causas, consequências, diagnóstico e gestão da imobilidade em doentes geriátricos, centrando-se em intervenções baseadas na evidência e na implementação de planos de cuidados personalizados.

a. Introdução teórica (Palestra e vídeo)

Conteúdo: Fornecer uma base teórica sólida sobre o que é a imobilidade, as suas causas mais frequentes na população idosa (doenças neurológicas, músculo-esqueléticas, etc.) e as complicações associadas, tais como úlceras de pressão, atrofia muscular e problemas psicológicos como a depressão.

Recursos:

- Leitura: Artigos e capítulos de textos sobre a epidemiologia e a fisiopatologia da imobilidade.
- Vídeo: Apresentações de especialistas que discutem casos e exploram a investigação mais recente sobre o tema.

Atividade:

b. Debate em linha

Fórum de discussão na plataforma de aprendizagem onde os estudantes podem colocar questões, partilhar experiências clínicas e discutir as melhores práticas para prevenir a imobilidade.

Tópicos para debate:

Estratégias para identificar doentes em risco, medidas preventivas eficazes em diferentes contextos de cuidados e a importância do trabalho interdisciplinar.

c. Estudo de caso

Descrição: Análise do caso de um paciente idoso com imobilidade grave devido a uma combinação de artrite grave e acidente vascular cerebral. Os alunos irão avaliar o paciente, identificar potenciais problemas e desenvolver um plano de gestão.

Actividades:

- Avaliação funcional utilizando instrumentos normalizados, como a Escala de Barthel.
- Conceção de um plano de reabilitação que inclua fisioterapia, intervenções nutricionais e apoio psicossocial.

d. Simulação virtual

Objetivo: Utilizar simulações na plataforma B-learning para praticar a avaliação física, o diagnóstico diferencial e a implementação de um plano de tratamento multidisciplinar.

Cenários simulados:

- Praticar a gestão de um doente que tenha desenvolvido imobilidade após uma cirurgia à anca.

- Intervenções para prevenir a imobilidade num doente com múltiplas comorbilidades.
Métodos:

e. Avaliação

- Questionário interativo: Perguntas de escolha múltipla e verdadeiro/falso para avaliar os
conhecimentos teóricos adquiridos sobre a imobilidade.
- Reflexão pessoal: Um relatório reflexivo em que o formando discute um plano de melhoria
contínua com base num cenário de caso fornecido, reflectindo sobre a forma como
aplicará os conhecimentos e competências aprendidos num ambiente clínico real.

f.Recursos adicionais

- Acesso a artigos de investigação actuais, para que os estudantes possam manter-se a par
dos últimos desenvolvimentos no tratamento e gestão da imobilidade.
- Webinars e conferências: Ligações para conferências em linha com especialistas em
geriatria e reabilitação para expandir a sua aprendizagem e compreender diferentes
perspectivas e abordagens.

2. Sequência didática detalhada para a síndrome de queda

Objetivo: Formar os médicos residentes para identificar os factores de risco de doença
cardíaca, avaliar eficazmente os doentes em risco e aplicar estratégias preventivas e
tratamentos eficazes na população geriátrica.

a. Apresentação interactiva (Webinar)

Conteúdo: Introdução à epidemiologia da caudaemia nos idosos, factores de risco
intrínsecos e extrínsecos, e as mais recentes orientações de prática clínica para a prevenção
da caudaemia.

Recursos:

- Webinar: Apresentação em direto por um especialista em geriatria que também responderá
às perguntas dos residentes em tempo real.
- Diapositivos interactivos: Apresentações que os alunos podem rever ao seu próprio ritmo
com ligações para estudos importantes e recursos adicionais.

b. Oficina virtual

Atividade: Sessões práticas em que os residentes aprendem a utilizar instrumentos de
avaliação do risco do café, como a Escala de Tinetti, o Teste de Levantar e Andar e
outras medidas de equilíbrio e mobilidade.

Simulação: Utilização de casos clínicos virtuais para praticar a avaliação de doentes, interpretar resultados e tomar decisões sobre intervenções adequadas.

c. Discussão em grupo

Descrição: Discussões em pequenos grupos utilizando plataformas de videoconferência para explorar e discutir diferentes estratégias preventivas adaptadas a vários contextos (como o hospital, o lar ou as instituições de longa duração).

Objectivos: Cada grupo deve desenvolver um plano de prevenção do café, tendo em conta as características específicas do ambiente e do paciente.

d. Jogo de papéis (Simulação)

Objetivo: Praticar intervenções em cenários controlados com actores ou simuladores que representem doentes idosos com elevado risco de café.

Actividades: Implementação de intervenções preventivas, utilização de dispositivos de assistência e como educar os doentes e os prestadores de cuidados sobre a prevenção de quedas.

Métodos:

e. Avaliação

- Prova prática de competências: Avaliação das competências adquiridas através de um circuito de estações onde os residentes demonstram as suas capacidades de avaliação e gestão dos riscos de queda de água.

- Teste de conhecimentos teóricos: Teste em linha que inclui perguntas de escolha múltipla, verdadeiro/falso e de resposta curta sobre a teoria da prevenção do café.

f.Recursos adicionais

Acesso a simulações avançadas: Uma plataforma que oferece cenários virtuais mais complexos para os residentes praticarem a identificação e o tratamento do café numa variedade de situações.

Leitura adicional e vídeos: Materiais que abordam estudos de caso, análises de intervenções eficazes e novas tecnologias na prevenção da catestesia.

3. Sequência Didática Detalhada para a Incontinência Urinária

Objectivos: Aprofundar a compreensão, o diagnóstico e a gestão da incontinência urinária na população geriátrica, promovendo uma abordagem integrada e multidisciplinar do tratamento.

a. Seminário

Conteúdo: Visão geral da incontinência urinária, incluindo a classificação dos diferentes

tipos (de esforço, de urgência, mista e de extravasamento) e revisão das opções de
tratamento actuais.

Recursos:

- Videoconferência: Apresentação de um especialista em urologia ou geriatria que explica
as bases fisiológicas e as implicações clínicas da incontinência.
- Artigos: Leituras atribuídas que abrangem estudos recentes e directrizes de prática
clínica actualizadas.
b. Trabalho em colaboração

Atividade: Em pequenos grupos, os residentes desenvolverão um plano de cuidados
interdisciplinares para um estudo de caso de um doente geriátrico com incontinência urinária,
tendo em conta os aspectos médicos, psicológicos e sociais.

Objectivos: Integrar os conhecimentos teóricos com a prática clínica, incentivando a
colaboração entre disciplinas como a urologia, a enfermagem, a fisioterapia e o serviço social.

c. Simulação clínica

Descrição: Exercícios práticos em plataformas de simulação que permitem aos residentes
efetuar avaliações de diagnóstico (tais como histórias clínicas, testes de esforço e
cistogramas), prescrever tratamentos e gerir o acompanhamento.

Actividades: Simulação de interacções com o paciente e decisões clínicas baseadas em
resultados de testes e respostas a tratamentos anteriores.

d. Fórum de discussão

Objetivo: Facilitar um espaço para a troca de experiências e estratégias sobre a gestão da
incontinência em diferentes contextos geriátricos.

Metodologia: Discussões facilitadas conduzidas por um moderador que coloca cenários
clínicos, dilemas éticos e questões sobre decisões de tratamento, encorajando os
residentes a contribuírem com as suas opiniões e aprendizagem prévia.

Métodos:
e. Avaliação

- Exame em linha: Teste que inclui perguntas de escolha múltipla, estudos de casos e
perguntas de redação para avaliar os conhecimentos teóricos e a capacidade de os aplicar
na prática.
- Apresentação de casos clínicos: Os residentes apresentarão uma análise detalhada de um
paciente fictício, incluindo avaliação, opções de tratamento e planos de acompanhamento,
demonstrando a sua capacidade de integrar e aplicar os conhecimentos adquiridos.

f.Recursos adicionais

Plataforma de aprendizagem: Acesso contínuo a uma plataforma com vídeos educativos, simulações interactivas e fóruns de discussão para reforçar a aprendizagem autónoma e contínua sobre a incontinência urinária.

Webinars e workshops: Oportunidades para participar em webinars e workshops práticos sobre as últimas inovações no diagnóstico e tratamento da incontinência urinária.

4. Sequência didática pormenorizada para o défice cognitivo

Objetivo: Dotar os médicos residentes de competências e conhecimentos críticos para avaliar, diagnosticar e gerir o défice cognitivo na população geriátrica, com uma abordagem abrangente que inclua intervenções médicas, psicológicas e sociais.

a. Curso em linha

Conteúdo: Uma introdução abrangente aos tipos de défice cognitivo, incluindo o défice cognitivo ligeiro (MCI) e várias formas de demência, como a doença de Alzheimer e a demência vascular.

Recursos:

- Vídeos educativos: Apresentações gravadas por especialistas em neurologia e geriatria, que discutem a base patológica e as manifestações clínicas do défice cognitivo.
- Leituras digitais: Artigos e capítulos de livros sobre as últimas investigações e tratamentos disponíveis.

b. Debate em linha

Atividade: Fóruns de discussão moderados onde os residentes analisam estudos de caso para identificar sinais e sintomas de deficiência cognitiva, discutir estratégias de avaliação e partilhar abordagens de gestão. Objectivos: Incentivar a reflexão crítica sobre os desafios diagnósticos e terapêuticos e melhorar a capacidade dos residentes para trabalhar numa equipa interdisciplinar.

c. Actividades interactivas

Descrição: Utilização de software especializado para efetuar avaliações cognitivas virtuais, incluindo o Mini-Mental State Examination (MMSE) e o Montreal Cognitive Assessment (MoCA). Prática: Simulações que permitem aos residentes aplicar estes instrumentos em cenários clínicos simulados, seguidas de feedback automático e explicações pormenorizadas sobre as respostas correctas.

d. Grupo de estudo

Actividades: Reuniões regulares do grupo de estudo para rever artigos recentes e discutir inovações no tratamento e gestão do défice cognitivo, incluindo terapias farmacológicas e não farmacológicas. Metodologia: Análise em pequenos grupos da literatura recente, com apresentações em grupo que resumem os principais resultados e a sua aplicabilidade clínica.

Métodos:
e. Avaliação

- Apresentação de um plano de gestão: Cada residente apresenta um caso completo, desde a avaliação inicial até ao plano de gestão, incluindo estratégias para gerir os aspectos médicos, psicológicos e sociais dos cuidados.
- Teste interativo: Um exame composto por perguntas de escolha múltipla e de resposta curta e por cenários de estudo de casos para avaliar os conhecimentos teóricos e as competências técnicas.

f.Recursos adicionais

Acesso a conferências e webinars: Ligações para conferências virtuais e webinars ministrados por especialistas em deficiência cognitiva e demência.

Ferramentas de aprendizagem ao longo da vida: Assinaturas de revistas de neurologia e geriatria e acesso a bases de dados de investigação para se manter atualizado sobre os desenvolvimentos na área.

5. Sequência de ensino pormenorizada para a sarcopénia

Objetivo: Dotar os médicos residentes de conhecimentos e competências práticas para identificar, prevenir e tratar a sarcopénia na população geriátrica, utilizando uma abordagem integrada que vai desde a avaliação à intervenção multidisciplinar.

a. Aulas multimédia

Conteúdo: Uma introdução detalhada à sarcopenia, incluindo a sua definição, mecanismos fisiopatológicos, impacto clínico e critérios de diagnóstico actuais de acordo com o consenso europeu (EWGSOP2).

Recursos:

- Vídeos educativos: Apresentações de especialistas em geriatria e fisioterapia que discutem os aspectos fisiológicos e as consequências da sarcopénia.
- Artigos e guias: Leituras digitais que fornecem informações actualizadas sobre os últimos desenvolvimentos na investigação e tratamento da sarcopenia.

b. Workshops práticos

Atividade: Sessões de formação sobre como implementar programas de exercício físico e estratégias nutricionais para prevenir e tratar a sarcopénia. Inclui formação sobre a utilização de equipamento de resistência e técnicas de treino de força.

Metodologia: Aprendizagem prática com demonstrações ao vivo e oportunidades para os residentes praticarem e receberem feedback direto.

c. Discussão em grupo

Objetivo: Analisar casos clínicos complexos para identificar a sarcopénia e discutir estratégias de intervenção multidisciplinar.

Formato: Discussões facilitadas em pequenos grupos, apoiadas por estudos de casos que reflectem diferentes cenários e níveis de gravidade da sarcopenia.

d. Laboratório virtual

Descrição: Simulações interactivas em que os residentes utilizam ferramentas de diagnóstico virtuais, como a bioimpedância eléctrica e o DEXA, para medir a composição corporal e diagnosticar a sarcopenia.

Prática: Exercícios online que permitem aos residentes interpretar os resultados dos testes e tomar decisões clínicas com base em cenários simulados.

e. Avaliação

Métodos:

- Exame prático e teórico: Avaliação das competências e dos conhecimentos adquiridos através de um exame que inclui questões de escolha múltipla e uma componente prática, em que os residentes devem demonstrar a sua capacidade de avaliar e planear tratamentos para doentes fictícios.
- Apresentação do projeto: Os residentes desenvolvem e apresentam um plano de intervenção abrangente para um caso de sarcopenia, considerando aspectos como o exercício, a nutrição e a gestão médica.

f. Recursos adicionais

Webinars e workshops online: Acesso a eventos educativos que apresentam inovações recentes e estudos de investigação sobre o tratamento e gestão da sarcopenia.

Plataforma de recursos em curso: Subscrição de uma plataforma que oferece actualizações regulares, vídeos instrutivos e artigos de interesse sobre a sarcopenia e a saúde geriátrica em geral.

6. Sequência didática pormenorizada para o
Síndrome de fragilidade nos idosos

Objetivo: Proporcionar aos médicos residentes uma formação abrangente na identificação, avaliação e gestão da fragilidade em adultos mais velhos, integrando abordagens multidisciplinares que abrangem tanto a prevenção como a intervenção.

a. Curso Introdutório Online

Conteúdo: Fundamentos da síndrome da fragilidade, incluindo a sua definição, critérios de diagnóstico e impacto na saúde e bem-estar dos idosos.

Recursos:

- Vídeos educativos: Série de vídeos que detalham a fisiopatologia da fragilidade, os factores de risco associados e as consequências clínicas.
- Leitura recomendada: Artigos e capítulos de livros sobre os modelos actuais de avaliação da fragilidade e a sua relevância clínica.

b. Workshops interactivos

Atividade: Workshops práticos sobre a utilização de instrumentos de avaliação da fragilidade, como a Escala de Fragilidade de Fried e o Índice Cumulativo de Fragilidade.

Metodologia: Aprendizagem prática com simulações de casos, em que os residentes aplicam estas ferramentas em pacientes idosos simulados, seguida de discussões em grupo sobre os resultados.

c. Fóruns de discussão

Objetivo: Discutir a gestão interdisciplinar da fragilidade, partilhando estratégias de intervenção de diferentes disciplinas (medicina, enfermagem, fisioterapia, serviço social)
Formato: Fóruns em linha onde os residentes podem trocar ideias, discutir estratégias de gestão e aprender com as experiências dos seus colegas e professores.

d. Simulações de gestão de casos
Descrição: Utilização de casos clínicos virtuais para praticar a tomada de decisão na gestão da fragilidade, incluindo o planeamento de intervenções como programas de exercício, modificações nutricionais e apoio psicossocial.

Práticas: Os residentes trabalham em equipas para desenvolver planos de gestão integrados, avaliando a eficácia de diferentes intervenções através de simulações interactivas.

Métodos:
e. Avaliação

- Avaliação prática: Os residentes apresentam os seus planos de gestão através de simulações de role-playing, recebendo feedback em tempo real sobre a sua abordagem clínica e capacidades de comunicação.
- Prova de conhecimentos: Prova escrita que abrange todos os aspectos teóricos da síndrome de fragilidade, garantindo a compreensão e a capacidade de aplicar os conhecimentos adquiridos.

f. Projeto de Melhoria Contínua

Atividade final: Desenvolvimento de um projeto de melhoria da qualidade num contexto geriátrico, em que os residentes identificam um problema relacionado com a fragilidade, analisam dados e implementam um plano de melhoria baseado em provas.

Objetivo: Aplicar os conhecimentos e as competências adquiridas num contexto real, a fim de melhorar os resultados dos doentes em contextos de cuidados geriátricos.
g. Recursos adicionais

Acesso a webinars e conferências: Ligações para conferências e webinars de especialistas em geriatria que debatem os últimos avanços e estudos na gestão da fragilidade.

Biblioteca digital: Acesso contínuo a uma biblioteca de recursos digitais com materiais de leitura, vídeos instrutivos e guias de boas práticas sobre fragilidade.

CAPÍTULO 5
UTILIZAÇÃO DA INTELIGÊNCIA ARTIFICIAL NA PRÁTICA GERIÁTRICA.

Introdução

O envelhecimento da população é um dos desafios mais significativos que a sociedade global enfrenta no século XXI. Na América Latina, esta tendência demográfica tornou-se uma realidade inegável, onde o aumento da esperança de vida e a diminuição das taxas de natalidade resultaram num crescimento constante da população idosa. Segundo dados da Comissão Económica para a América Latina e as Caraíbas (CEPAL), prevê-se que, em 2024, mais de 15% da população total da América Latina tenha 65 anos de idade ou mais. Esta mudança demográfica coloca desafios significativos para os cuidados de saúde, a qualidade de vida e o bem-estar desta população crescente de adultos mais velhos.

Neste contexto, a tecnologia, e em particular a Inteligência Artificial (IA), surgiu como uma ferramenta promissora para enfrentar os desafios e as necessidades dos cuidados de saúde geriátricos. A IA tem o potencial de transformar radicalmente a prática médica, oferecendo soluções inovadoras e personalizadas que podem melhorar a qualidade de vida dos idosos, otimizar a gestão dos recursos no sistema de saúde e apoiar os profissionais de saúde na tomada de decisões clínicas.

À medida que nos aproximamos de um futuro em que os cuidados de saúde para adultos mais velhos desempenham um papel central no desenvolvimento sustentável da América Latina, é essencial explorar e compreender o papel da IA neste contexto. Esta investigação visa lançar luz sobre os desafios e oportunidades que a IA apresenta contribuindo assim para o avanço da medicina geriátrica na região e para o bem-estar dos seus cidadãos seniores.

1. Inteligência artificial aplicada aos cuidados geriátricos

a. Definição de Inteligência Artificial

A inteligência artificial (IA) é um domínio rapidamente emergente na tecnologia educativa, incluindo no contexto da educação médica (Zawacki-Richter O ;et al, 2019). A IA, impulsionada por algoritmos de aprendizagem automática, está a ganhar popularidade no sector dos cuidados de saúde e tem o potencial de melhorar os cuidados aos doentes, a análise de dados em tempo real e a monitorização contínua dos doentes (Kolachalama, V. e Garg, P. , 2018) (Sapci, A. e Sapci, H. , 2020).

A IA refere-se ao desenvolvimento de sistemas e algoritmos capazes de realizar tarefas que normalmente requerem inteligência humana, como o processamento de linguagem natural, a aprendizagem automática e a tomada de decisões.

b. A.I. em Educação Médica.

No domínio da educação médica, a IA tem diversas aplicações, como a aprendizagem personalizada, a análise de dados, o reconhecimento de imagens médicas e a tomada de decisões sobre planos de tratamento (Chan, K. e Zary, N., 2019) (Xu, H: et al, 2022). No entanto, a adoção da IA na educação médica enfrenta desafios, incluindo a necessidade de os educadores estarem mais envolvidos na investigação e implementação da IA (Zawacki-Richter O ;et al, 2019).A integração da IA nos currículos das escolas médicas já está a ser explorada, e as aplicações da tecnologia de IA estão a ser sentidas numa variedade de disciplinas médicas (Wood, E; et al, 2021). Há um reconhecimento crescente do potencial da IA na educação médica, incluindo a sua capacidade de melhorar os processos de ensino e aprendizagem (Popenici, S. e Kerr, S. , 2017). No entanto, a adoção da IA no ensino médico exige uma análise cuidadosa de fatores como a revisão dos programas curriculares, a inclusão de conceitos básicos de IA no ensino médico e o desenvolvimento de ferramentas baseadas em IA (Iqbal, S., 2022) (Memon, S;et al, 2021). A utilização da IA no ensino médico também suscita considerações éticas e a necessidade de formação em bioética (Briganti, G. e Moine, O., 2020). Um aspeto fundamental da IA no contexto da educação médica é a sua capacidade de fornecer soluções personalizadas e adaptativas. Neste sentido, é importante reconhecer que a incorporação da inteligência artificial na área médica, a partir da dimensão profissional e académica, não se destina a substituir o trabalho ou o conhecimento humano, mas sim a apoiar a prática médica e a fornecer as ferramentas de aprendizagem necessárias ao sujeito, facilitando a sua compreensão durante o processo de formação (Aguilar Bucheli, D; et al, 2023).Os algoritmos de aprendizagem automática (ML), um ramo da IA, estão a ganhar popularidade no sector da saúde. A IA alimentada por algoritmos de aprendizagem automática tem o potencial de transformar a educação médica, proporcionando experiências de aprendizagem personalizadas e melhorando a precisão do diagnóstico (Kolachalama, V. B. & Garg, P. S., 2018). A inteligência artificial (IA) tem demonstrado um enorme potencial para transformar a prestação e a acessibilidade dos cuidados de saúde na América Latina. A utilização da IA neste contexto pode melhorar a saúde e o bem-estar dos idosos, ajudar nos cuidados de enfermagem e enfrentar os desafios relacionados com a adesão à medicação e a gestão das doenças (Garcfa Alonso R; et al, 2022).

c. Situação da IA na América Latina

Na América Latina, os sistemas de saúde variam de um país para outro e encontram a sua principal referência de aplicação no domínio dos cuidados oncológicos (Garcfa Alonso R; et al, 2022) (Liliana, Sussman; et al, 2022); a IA tornou-se importante nos cuidados oncológicos, com resultados promissores na previsão de parâmetros clinicamente relevantes, no diagnóstico do cancro, na investigação e na medicina personalizada. No entanto, o desenvolvimento e a cooperação em pesquisa de IA e cuidados geriátricos na América Latina são limitados.Fortalecer a colaboração e a comunicação entre países, regiões e instituições pode impulsionar ainda mais o desenvolvimento da IA em cuidados geriátricos na América Latina [3].Entre os temas e áreas de estudo em nossa região podemos citar:Melhoria da Atenção à Saúde do Idoso: O uso da IA tem um impacto na melhoria da segurança do paciente nos cuidados de saúde. Isto inclui a deteção precoce de doenças como a demência e outros problemas geriátricos, a identificação de reacções adversas a medicamentos durante a hospitalização e a criação de listas de reconciliação de medicamentos para reduzir os erros clínicos e melhorar o envelhecimento (Jehath Syed, 2022).Oportunidades para a prática de enfermagem gerontológica: Foi discutido o modo como a IA pode fazer avançar a prática de enfermagem gerontológica, embora não estivessem disponíveis pormenores específicos do estudo, este realça a importância da IA na saúde dos adultos mais velhos e o seu potencial para melhorar a prática gerontológica (O'Connor S, 2022). Revisão da utilização da IA nos cuidados aos idosos: Um estudo de revisão abrange vários tipos de tecnologias de IA utilizadas nos cuidados a idosos, tais como robôs, dispositivos ex-esqueléticos, casas inteligentes, aplicações de saúde inteligentes e dispositivos activados por voz. Estas tecnologias desempenham papéis como terapeutas de reabilitação, apoios emocionais, facilitadores sociais, supervisores e promotores cognitivos. O impacto da IA nos cuidados aos idosos é promissor, embora seja necessária mais investigação para validar estas funções (Ma, B. et al, 2023).

Impacto da IA na saúde dos idosos e potencial discriminação: A Organização Mundial de Saúde salientou a capacidade da IA para prever riscos para a saúde e personalizar a gestão dos cuidados de saúde. No entanto, também alerta para o risco de discriminação em função da idade devido a enviesamentos nos dados que alimentam estas tecnologias e na conceção das mesmas. Foram propostas políticas para garantir que a IA tem um impacto positivo na vida dos idosos e que a discriminação em razão da idade é evitada (ONU, 2022).

Equilíbrio entre segurança e autonomia para os adultos mais velhos: Um workshop explorou a IA no contexto do equilíbrio entre a segurança e a autonomia dos adultos mais velhos e das pessoas com deficiência. Esta abordagem sublinha a importância da IA para ajudar estas populações a viver de forma tão independente quanto possível (Lustig, T. A., & Cilio, C. M., 2019).

IA nos cuidados a idosos que vivem sozinhos: Um estudo realizado em Espanha demonstra como a IA pode ajudar no cuidado de pessoas idosas que vivem sozinhas, permitindo que os cuidadores familiares monitorizem através de uma aplicação móvel. Esta abordagem aumenta a segurança e o bem-estar do membro da família e inclui também serviços de companhia remota e assistência técnica (Infogeriatria, 2022). A relevância desta investigação reside na necessidade imperativa de explorar as características da aplicação de tecnologias virtuais por inteligência artificial nos cuidados geriátricos e gerontológicos na América Latina. A relevância deste estudo reside na necessidade urgente de compreender como a IA está a ser implementada e utilizada na prática médica geriátrica no contexto latino-americano no ano de 2024. Através de uma abordagem positivista e de um desenho exploratório, este estudo pretende esclarecer as seguintes questões:

1. Avaliação da adoção e utilização de tecnologias de IA nos cuidados a idosos na América Latina.

2. Identificação das vantagens e dos desafios associados à integração da IA na prática médica geriátrica na região.

3. Análise dos impactos potenciais da IA na qualidade de vida dos idosos e na eficiência dos serviços de saúde.

4. Avaliação das perspectivas e atitudes dos profissionais de saúde geriátricos em relação à IA como ferramenta complementar na sua prática.

5. Identificação de possíveis recomendações para a melhoria da implementação e adoção da IA nos cuidados de saúde geriátricos na América Latina.

Neste sentido, propõe-se a realização de um estudo exploratório para estabelecer as bases de futuras investigações, em relação às características da utilização da inteligência artificial na prática médica geriátrica nos países membros do COMLAT (Comité Latino-Americano de Geriatria).

O estudo de Garcia Alonso (2022) (Garcfa Alonso R; et al, 2022) conclui que: A inteligência artificial (IA) tem o potencial de transformar a prestação e a acessibilidade dos cuidados de saúde na América Latina, particularmente nos países de baixo e médio rendimento.

A inteligência artificial é também utilizada para a análise preditiva nos cuidados de saúde, o que ajuda a identificar os doentes em risco de desenvolver determinadas doenças ou complicações. Este facto pode ajudar a uma intervenção precoce e a planos de tratamento personalizados.

Os chatbots e os assistentes virtuais com inteligência artificial são utilizados para fornecer informações básicas de saúde e apoio aos doentes, ajudando a melhorar o acesso aos serviços de saúde.

A IA está a ser utilizada para a descoberta e o desenvolvimento de medicamentos, ajudando a acelerar o processo de identificação de potenciais novos medicamentos e tratamentos. A inteligência artificial é utilizada na monitorização remota de doentes, permitindo aos prestadores de cuidados de saúde monitorizar remotamente os sinais vitais e o estado de saúde dos doentes, permitindo intervenções atempadas e reduzindo a necessidade de visitas presenciais.

O estudo de Jingjing Wang, 2023 (Wang J, et al, 2023) "Application of artificial intelligence in geriatric care: bibliometric analysis" (Aplicação da inteligência artificial nos cuidados geriátricos: análise bibliométrica) analisa os actuais pontos críticos de investigação e as redes de colaboração na aplicação da IA nos cuidados geriátricos através da análise bibliométrica.

O estudo revelou que a investigação sobre a aplicação da IA nos cuidados geriátricos se desenvolveu rapidamente, com um aumento significativo de publicações entre 2014 e 2022, representando 90,87% de todas as publicações.

Os principais pontos críticos de investigação identificados neste domínio incluem a doença de Alzheimer, os cuidados aos idosos, a aceitação e a monitorização e tratamento de doenças.

A aprendizagem automática, a aprendizagem profunda e a reabilitação tornaram-se recentemente pontos críticos da investigação sobre a aplicação da IA nos cuidados geriátricos.

Os Estados Unidos e o International Journal of Social Robotics foram os principais contribuintes em termos de número de publicações sobre este tema.

2. Considerações Considerações éticas

Questões de privacidade e transparência no uso de dados e registos de pacientes, bem como dificuldades tecnológicas e regulatórias, são desafios enfrentados pelos países latino-americanos na implementação de serviços baseados em inteligência artificial nos cuidados de saúde. A prestação de serviços de saúde baseados em inteligência artificial na América Latina contribui para a promoção dos Objectivos de Desenvolvimento Sustentável das Nações Unidas (Garcfa Alonso R; et al, 2022).

REFERÊNCIAS

• Garcfa Alonso R; et al. (2022). Saúde Digital e Inteligência Artificial: Avanço da prestação de cuidados de saúde na América Latina. Profissional de TI. doi:doi: 10.1109/mitp.2022.3143530

• Aguilar Bucheli, D; et al. (2023). Inteligência artificial na educação médica latino latino-americano contexto...Metro Ciencia, https://doi.org/10.47464/metrociencia/vol31/2/2023/21-34.

• Briganti, G. e Moine, O. (2020). Inteligência artificial em medicina: hoje e amanhã. Frontiers inMedicine, 7, https://doi.org/10.3389/fmed.2020.00027.

• Chan, K. e Zary, N. (2019). Aplicações e desafios da implementação da inteligência artificial na educação médica: revisão integrativa. . Jmir Educação Médica, https://doi.org/10.2196/13930.

• Doumat, G;et al. (2022). Conhecimento e atitudes dos estudantes de medicina no Líbano em relação à inteligência artificial: um estudo de inquérito nacional. Frontiers in Artificial Intelligence, https://doi.org/10.3389/frai.2022.1015418.

• Dziatkovskii, A. (2023). O efeito ergonómico da IA & ML na educação. https://doi.org/10.46916/26042023-1-978-5-00174-960-8.

• Infogeriatria (2022). Recuperado em 01 02 02 2023, de Infogeriatria: https://www.infogeriatria.com/noticias/20220113/estudio-demuestra-Inteligência artificial ajuda a cuidar dos idosos que vivem sozinhos

• Iqbal, S. (2022). Os educadores médicos estão preparados para adotar a inteligência artificial no sistema de saúde e na educação médica? Health Professions Educator Journal, 7-8. doi:https://doi.org/10.53708/hpej.v5i1.1707

• Jehath Syed (2022). Recuperado em 02 02 2024, de s4be.cochrane.org: https://s4be.cochrane.org/blog/2022/10/14/involving-artificial-tecnologias-da-informacao-no-cuidado-dos-idosos-o-futuro-dos-saude/

• Kolachalama, V. e Garg, P. (2018). Aprendizagem automática e educação médica. . NPJ Medicina Digital, https://doi.org/10.1038/s41746-018-0061-1.

• Kolachalama, V. B. & Garg, P. S. (2018). Aprendizagem de máquina e educação médica. , 1(1). . NPJ Digital Medicine, https://doi.org/10.1038/s41746- 018-0061-1.

• Liliana, Sussman. et al. (2022). Integração de inteligência artificial e oncologia de precisão na América Latina. Fronteiras em tecnologia médica. doi:doi: 10.3389/fmedt.2022.1007822

• Lustig, T. A., & Cilio, C. M. (2019). Aplicações de Inteligência Artificial para Idosos e Pessoas com Deficiência: Equilíbrio entre segurança e autonomia: Proceedings of a Workshop-in Brief. (E. a. Academias Nacionais de Ciências, D. o. Education, H. a. Division, B. o. Services, B. o. Policy, & D. a. Forum on Aging, Edits). National Academies Press (EUA). doi:DOI: 10.17226/25427

• Ma, B. et al. (2023). Inteligência artificial nos cuidados de saúde dos idosos: A scoping review.Revisões da investigação sobre o envelhecimento. doi:https://doi.org/10.1016/j.arr.2022.101808

• Memon, S;et al. (2021). Perceção sobre inteligência artificial na educação médica. PJMHS,419-420.doi: https://doi.org/10.53350/pjmhs2023173419

• O'Connor S. (2022). Inteligência artificial para a saúde do adulto mais velho: Oportunidades para o avanço da prática de enfermagem gerontológica. Journal of gerontological nursing, 48(12), 3-5. doi:https://doi.org/10.3928/00989134-20221107-01

• ONU. (2022). Recuperado em 01 02 2023, de UN News: https://news.un.org/es/story/2022/02/1503842

• Popenici, S. e Kerr, S. (2017). Explorar o impacto da inteligência artificial no ensino e na aprendizagem no ensino superior. . Investigação e Prática em Aprendizagem Melhorada pela Tecnologia, https://doi.org/10.1186/s41039-017-0062- 8.

• Sapci, A. e Sapci, H. (2020). Educação e ferramentas de inteligência artificial para estudantes de informática médica e de saúde: revisão sistemática. . Jmir Educação Médica, https://doi.org/10.2196/19285.

• Wang J, et al. (2023). Aplicação de Inteligência Artificial na Análise Bibliométrica de Cuidados Geriátricos. J Med. doi:DOI: 10.2196/46014

• Wood, E; et al. (2021). Estamos prontos para integrar a alfabetização em inteligência artificial no currículo da escola de medicina: pesquisa com alunos e professores. . Journal of Medical Education and CurricularDevelopment, https://doi.org/10.1177/23821205211024078.

• Xu, H: et al. (2022). Caminho de cultivo de talentos compostos em diagnóstico oftálmico, tratamento e enfermagem com base na inteligência artificial. Journal of Clinical and Nursing Research, 106-111. doi:https://doi.org/10.26689/jcnr.v6i5.4387.

• Zawacki-Richter O ;et al. (2019). Revisão sistemática da investigação sobre aplicações de inteligência artificial no ensino superior - onde estão os educadores? Revista Internacional de Tecnologia Educacional no Ensino Superior, https://doi.org/10.1186/s41239-019-0171-0.

ÍNDICE

Printed by Books on Demand GmbH, Norderstedt / Germany